AGRARIAN TRANSFORMATION IN TRIBAL AREAS

EMERGING TRENDS AND ISSUES

AGRARIAN TRANSFORMATION IN TRIBAL AREAS

EMERGING TRENDS AND ISSUES

By

Dr. Hari Charan Behera

Assistant Professor,
Centre for Rural Studies
LBS, National Academy of Administration
Mussoorie - 248 179

DISCOVERY PUBLISHING HOUSE PVT. LTD.
NEW DELHI-110 002

First Published-2010

ISBN 978-81-8356-636-0

Published by:

DISCOVERY PUBLISHING HOUSE PVT. LTD.
4831/24, Ansari Road, Prahlad Street
Darya Ganj, New Delhi-110002 (India)
Phone: +91-11-23279245, 43764432 • Fax: +91-11-23253475
***E-mail:* parul.wasan@gmail.com**
info@discoverypublishinggroup.com
***web:* www.discoverypublishinggroup.com**

Printed at:
Sachin Printers
Delhi

PREFACE

Agriculture continues to be the backbone of Indian economy with about 65% of the Indian population depends on agriculture directly and it accounts for 22% GDP. Notwithstanding the fact that agriculture is deeply attached to the livelihood of millions of people, the support from institutional service providers to the farmers, penetration of technological inputs and adoption and dissemination of technology in agriculture is skewed across geographical regions. In backward and hilly regions, where the tribal communities inhabit, agricultural issues remain to be prominent nearly after six decades of the plan period in India. The issues lie in both parts of institutional and technological dimensions. The consequent effort of the private and government agencies to overcome the issues of backwardness in agriculture has little or no effect on the communities in maximizing their benefits out of existing agricultural practices. The challenges in agriculture in tribal areas are multifarious and multidimensional. Multiple factors attached to the performance of agricultural operations and production activities in tribal areas. In addition to the socio-economic and ecological factors, culture also plays a significant role in agriculture among the indigenous tribal communities. The agricultural practices vary across geographical regions and communities with diverse local conditions. The government's effort to recognize the local conditions and formulate strategies for better extension in agriculture continues to be a big challenge.

Despite the fact that there is a rapid change in cropping pattern and technological adoption in some rural areas, in the tribal belt, the change is rather slow or stagnant. The tribal farmers face more challenges in accessing necessary technological inputs, financial inputs, knowledge inputs, and any other supports from both private and public agencies. The new economic policies, on the other hand, have affected both traditional agricultural practices and the new agricultural practices that were introduced during the post-green revolution period.

This book is an effort to give a systematic focus on agrarian transformation in tribal areas with the support of fieldwork in Kandhamal district, a tribal dominated district, in Orissa. A systematic anthropological approach of research design and field survey has helped the researcher in generating clues about the issues persisting in agriculture in the tribal belts. A dynamic approach of understanding the issues from a close quarter is the strength of the book.

HARI CHARAN BEHERA

ACKNOWLEDGEMENTS

I express my deep gratitude to my revered teacher and supervisor, Professor P. Venkata Rao, who has been instrumental in shaping the study in the present form. The present study would not have been possible but for his relentless effort, sincerlty and astute interventions at its various stages. His subtle anthropological approaches obviously very often force me to look at the issues involved afresh.

I am thankful to Professor P. K. Nayak, Head of the Department of Anthropology, Utkal University providing basic inputs and suggestions for requisite corrections before bring out this book into this present form.

I am grateful to the members of the faculty in the Department of Anthropology, University of Hyderabad: Professor K.K. Misra, Professor N. Sudhakar Rao, Dr. R. Siva Prasad, Dr. B.V. Sharma, Dr. George Tharakan, Dr. Abdul Munaf and Dr. M. Romesh Singh, as they extended their helping hands whenever they were approached.

I am equally grateful to Professor Premananda Panda, Professor Deepak Kumar Behera, Professor P.N. Sahu and Dr. R. Choudhury at the Department of Anthropology, Sambalpur University for being a constant source of inspiration not only during research work but also beyond.

I am very thankful to Shri Ashish Vachhani, IAS, Coordinator, Centre for Rural Studies, LBSNAA for his academic inputs, encouragement and providing all possible support for preparation of this book.

I sincerely thank to my colleagues Dr. Saroj Arora, Ms. Melanie Hilton, Dr. V.V. Singh, Shri Ramesh Kothari, Shri Samar Kashyap and others at the Centre for Rural Studies, LBS, National Academy of Administration, Mussoorie for being cooperative throughout towards completing this work.

As a matter of fact, the study would have been out of my reach, had I not received the support and cooperation from the villagers (the respondents for this study) of the Kandhamal district in Orissa – starting from arranging my accommodation for a pretty long period of time, sharing crucial information about different contentious issues, being friendly all the time, and so on.

At length, my immense gratitude to my family members whose constant support and encouragement are the real forces that always keeps me away from disintegration.

HARI CHARAN BEHERA

CONTENTS

ABBREVIATIONS

ADAO	Additional District Agricultural Officer
ADR	Associate Director of Research
AEO	Agricultural Extension Officer
AICP	All-India Cost Price
AMCS	Agency Marketing Cooperative Society
APEDA	Agricultural and Processed Food Products Export Development Agency
APL	Above Poverty Line
AWW	Anganwadi Worker
BPL	Below Poverty Line
CBO	Community-based Organisation
CRRI	Central Rice Research Institute
CST	Central Sales Tax
DAO	District Agricultural Office
DPAP	Drought Prone Area Programme
DRDA	District Rural Development Agency
FAO	Food and Agricultural Organisation
FPR	Farming Participatory Research

FSR	Farming System Research
FTP	Foreign Trade Policy
FWP	Food for Work Programme
HDI	Human Development Index
HYV	High Yielding Variety
IK	Indigenous Knowledge
IKS	Indigenous Knowledge System
IMR	Infant Mortality Rate
IRDP	Integrated Rural Development Programme
ITDA	Integrated Tribal Development Agency
KASAM	Kandhamal Apex Spices Association for Marketing
KVK	Krishi Vigyan Kendra
LAMPCS	Large Size Multi Purpose Cooperative Societies
MFP	Minor Forest Produce
MNC	Multinational Company
NABARD	National Bank for Agricultural and Rural Development
NATP	National Agricultural Technology Project
NREGP	National Rural Employment Guarantee Programme
NTFP	Non-timber Forest Produce
OAP	Old Age Pension
OUAT	Orissa University of Agriculture and Technology
PESA	Panchayat Extension to Scheduled Areas Act
PMGSY	Pradhan Mantri Gram Sadak Yojna
PRA	Participatory Rural Appraisal
RMP	Regional Medical Practitioner
RRTTS	Regional Research Technology Transfer Station

SC	Scheduled Caste
SDS	Spices Development Society
SEZ	Special Economic Zone
SGSY	Swarna Jayanti Grameen Swarozgar Yojna
SHG	Self-Help Group
SMM	Semi Mechanical Method
ST	Scheduled Tribe
TDCC	Tribal Development Cooperative Cooperation
TKS	Traditional Knowledge System
TOT	Transfer of Technology
TRYSEM	Training of Rural Youth for Self Employment
UNDP	United Nation Development Programme
VAW	Village Agricultural Worker
VLW	Village Level Worker
WP	Widow Pension
WTO	World Trade Organisation

ABSTRACT

The title as suggests is an attempt to examine the emerging trends in agriculture and transformation that is taking place in cropping pattern, technology adoption and other associated dimensions in agriculture in tribal areas. The other important focus in the book is the analyses of the impact of institutional intervention and implication of new economic policies for agriculture in the said areas. The book comprises of seven chapters including introduction and conclusion. The chapter outlines are mentioned below:

Chapter I: Introduction

It includes basic understanding about the innovation, adoption and dissemination of technology, review of literature, conceptual and theoretical understanding, objectives of the study, research framework and chapter outlines.

Chapter II: Profile of the Villages

While describing profile of the villages, there is an attempt to sketch out the district profile, which includes geographical location, population, administration, health, literacy, etc. As there are two different social groups such as the Kandha and the Pana selected under the study, a brief ethnography on these two groups is provided in the chapter.

Physical features, economic organisation, social organisation, political organisation, education, health, religion, rituals, festivals beliefs, etc are discussed briefly under the ethnography of the Kandha. Similarly, a very brief discussion about the Panas is also presented in the chapter.

This chapter, 'Profile of the Villages' covers the geographical location of the study area, flora, fauna, demography, health and sanitation, social infrastructure, education, economy, etc.

Chapter III: Traditional Agriculture

This chapter deals with the detailed discussion about the traditional knowledge, innovation, and adoption of technology of the farmers in the region. The focus is on the types of agricultural practices, types of crops grown, tools and techniques used in the traditional agriculture, etc. A discussion about the podu cultivation and crops grown in the field is also included in the chapter. There is description about the settled agricultural practices in traditional agriculture surrounding crops selection, rationality and other practices including adoption of technology. How the traditional knowledge among tribal farmers helps in agricultural production has been dealt with in the chapter.

This chapter explores the innovation process that is carried by the tribal community in the traditional agriculture. This focuses on the local knowledge and agriculture, the indigenous mechanisms, the water management practices, the tools used in traditional agriculture, the rationality in adoption and the uses of technologies and other factors influencing the adoption process.

Chapter IV: Trends in Agriculture and Adoption of Technology

The fourth chapter deals with the changing aspects of agricultural practices and the adoption of technology. It describes the current agricultural practices of the Kandha and the Pana in the villages. The landholding detail, a precondition for agriculture, is discussed in the chapter. This covers the changing landholdings pattern in the villages over a period of time. Landholding through inheritance, land brought under possession, land under mortgage and share cropping are also discussed under the landholding details. Division of labour and role of women in agriculture is touched upon here. Variation in technology handling between men and women is also outlined briefly. Growth of vegetable cultivation in the villages is one of the positive trends is discussed in this chapter. It covers production strategy, use of tools, techniques, application of traditional and modern agricultural implements, knowledge and innovation, and dissemination of technology in agriculture. The chapter deals with both positive and negative aspects of the use of fertilisers and pesticides in agriculture. The choice, acceptance and reasons for acceptance of certain varieties of paddy and other crops form the important parts of the discussion. Some of the problems in the current agricultural practices are identified in the chapter. Moreover, this chapter also tries to discuss the existing traditional agricultural practices and their trends towards transformation.

Chapter V: Effectiveness of Involvement of Public and Private Agencies in Agriculture

This chapter critically examines the nature of intervention of formal sectors in agriculture. In addition, this chapter tries to find out the extent the agencies have delivered the services in agricultural development.

The chapter outlines the gap between the tribal needs and the provision of services by the public and the private agencies. The discussion is on the role of important agencies like Regional Research Technology Transfer Station, Krishi Vigyan Kendra, and Soil Conservation Department. Some of the important agricultural schemes and their effectiveness of operations are covered in the chapter.

The role of private agencies, cooperatives and NGOs is a significant part of the chapter. Some NGOs like KASAM and SAMANWITA, and Cooperative Agencies like OMFED are taken into consideration to examine their roles and effectiveness. In addition, the important discussion in the chapter is the role of grass root level workers like village agricultural worker (VAW) and village level worker (VLW) in extension of services for the agricultural development in the villages. There is focus on the issues with regard to extension and delivery of services in the present scenario. This chapter is an attempt to examine the role of the agencies in the communities. This chapter reflects some of the pros and cons of the intervention of the agencies in agriculture. There are some important issues pertaining to the role of agricultural officials, researchers, and extension officers, approach of the extension workers and work culture of the agencies in the region, which the author has outlined with the support of case studies.

Chapter VI: New Economic Policies and their Implications for Agriculture

This chapter is significant from the view point of understanding the impact of new economic polices on the agriculture in tribal areas. It primarily highlights the rural development schemes and programmes and the institutional reforms. Institutional reforms include credit reforms, tax reforms, privatization and electricity reforms. Privatization of electricity sector, seed and fertilizer sector constitute an important part of the discussion. The chapter highlights the consequences of privatization of seed supply in agriculture. Some other important points that are highlighted include agri-export zone, private entry into agri-production and business and their impact on agriculture in the region. The chapter explains various issues at the micro-level and explores their link between micro and macro policies.

Chapter VII: Conclusion

The conclusion chapter presents the gist of each chapter and highlights the farmers' needs as an approach for successful developmental programs that will be helpful for removing the existing drawbacks in agricultural development in the region.

CHAPTER - I

INTRODUCTION

I

Agriculture is the most significant discovery of human beings. Human civilization started with the development of agricultural practice and with the changes in technology necessitated by this activity. Managing agricultural activities requires both mental and physical abilities. It is an art. Performance of agricultural practice needs skills and techniques. Ever since the origin of agriculture in Neolithic period, man has constantly engaged himself in innovating in skills and techniques for further development in agriculture.

Anthropology in the course of its development continued the emphasis on peasant, tribal or rural societies. Agriculture has been an important means for the livelihood of millions of people. Anthropological interest in the study of agriculture began since the time of cultural evolutionists. Cultural evolutionists like Tyler, Morgan, and others have discussed about the courses of evolution of culture and the development of agriculture. Julian Steward (1955), while discussing about the cultural ecology, emphasized the agricultural zone and environmental adaptability. Gordon Childe (1951) discussed the origin of agriculture in the history of mankind. He pointed out how mankind began to change its mode of life from food gathering to food cultivation, by adopting shifting cultivation, which is historically a transitional stage towards more permanent agriculture. Leslie White (1959), in his theory of evolution, has indicated that culture evolves as the amount of energy harnessed per capita per year is increased, or as the efficiency of the instrumental means of putting the energy to work is increased, which in due course leads to agricultural growth. This cultural apparatus leads to socio-economic development of the community. Agriculture succeeded pastoralism according to the early evolutionists. Ratzel (1896) regarded that agriculture is an improvement because it forces on a man the wholesome habit of the labour and is followed by the accumulation of capital, development of trade, and fuller organization of social ranks. Thus modern (Europian) agriculture represented a further advancement over the insecure gardening of primitive agriculture among Negroes and Polynesians.

CULTURE, AGRICULTURE AND ENVIRONMENT

The socio-economic and ecological approaches in agriculture brought broadly a new vision in the anthropological discipline during twentieth century. Kroeber's (1939) view of environmental approach, Megger's (1954) continuity with 'possibilist approach and Steward's (1955) view of cultural ecology and environmental determinism, point to the idea of linkages of culture, agriculture and environment. Megger's possibilist approach divided environments into four types on the basis of their 'subsistence potential' for agriculture, and stated the law that the level to which a culture can develop is dependent upon the agricultural potentiality of the environment it occupies. His assertion that certain areas showing an agricultural potential is decisively limited by low soil fertility resulting from abundant rain fall and high humidity. Gupta (1995) from his researches in India and Bangladesh shows the importance of socio-ecological perspectives in agriculture. He has stressed the importance of understanding these dimensions on farmers' innovations and risk adjustments. Economic approach to agriculture is also quite common in anthropology. Many great anthropologists like Dalton, Firth, Sahlins and Epstein have substantial contribution to economic approach in agricultural development. In their approaches following questions such as, how does the change in economy lead to cultural change or vice versa, how the primitive economy is shaped by natural and social environment, how organization plays crucial role in agriculture are discussed and answered.

AGRARIAN TRANSFORMATION

The study of agrarian transformation is not new in anthropological studies. In India many anthropologists, and sociologists like Baily (1957), Epstein (1962) and Beteille (1974) have tried to analyse the Indian scenario through case studies. However, these anthropologists have followed different theoretical approaches to study on agrarian changes. But, there is similarity in their analysis in the context of understanding changes, may be social or economic. Baily's approach of caste and economic frontier analyses the relations between caste and tribe in a Phulbani highlander village in Orissa with respect to economy that includes agriculture and marketing. In acquisition of wealth and the transfer of land, he elaborately discussed about why does land come into the market. He has focused the socio-cultural and economic causes for the sale of land. Epstein (1962) traces out the agrarian changes through village studies in Karnataka. She explained how difference in access to irrigation leads to differences in economic development in two villages such as Wangala and Dalena. According to her the villagers are not slow to react to new economic opportunities. She supported Mead's (1954) observation about how "most of the farmers of the world are not motivated by abstract ends or speculative results. For them, seeing is believing" (1962, 311). Beteille (1974) in agrarian studies mainly focused the agrarian social structure aspect principally derived from landholding point of view.

In contrast to the above, the study of technological processes, for a longer period, has lacked specialized attention. Only a few anthropologists like Leslie White discussed the role of technology in cultural evolution. But, still the inadequacy lies on the study of technological aspects, particularly in the study of agriculture in anthropology. Only in recent years anthropologists have begun to challenge the exclusion of technology from culture and society, in connection with a general re-examination of the ways in which the key concepts that structure their enquiry have been shaped by the concept of western modernity. There is no doubt that the adoption of the concept of technology has had a profound effect on our thinking about the relation between human beings and their activity (Encyclopaedia of Cultural Anthropology, 1996). The work on technology has been substantially contributed by other disciplines such as Economics, Sociology, Agricultural Science, etc. But, understanding beliefs, rationalities and options in choosing certain practice in the form of crops and technologies received less attention.

II

CONCEPTUAL UNDERSTANDING

Understanding agrarian transformation is the interest of different subject specialists but in a different form. For agricultural scientists it is mainly the study of rigorous scientific and technological aspects such as agronomy, plant ecology, pathology, plant biotechnology, etc. For an economist, it is purely the study of economic aspects of production and marketing which analyses mainly profit and loss; for a sociologist it is mainly the study of social and behavioural understanding of agrarian aspects. The anthropologists on the other hand, are more or less close to sociology but it mainly analyses the agrarian aspect from holistic perspective. For, agrarian transformation is not only the understanding of changing relationship between landlords and peasants, caste and tribe or landowners and tenants. Agrarian transformation is a broader process encompassing the transformation of production of crops, use of tools, beliefs, rationalities, relationships, institutions, policies, etc. The analysis of multiple aspects is most essential for an anthropologist eager to see agrarian transformation from developmental angle.

Therefore, the following discussion focuses technological dimensions, institutional aspects and policy aspects for understanding agrarian transformation.

TECHNOLOGICAL DIMENSIONS

The word "technology" is formed by compounding the classical roots of *techno* meaning *'art'* in the traditional sense of craft or skill and *logos* meaning a framework of principles derived from the application of reason.

Pfaffenberger (1992) employed two definitions of technology. One is more restricted and the other is more inclusive. Technique refers to the system of material resources, tools, operational sequences and skills, verbal and non-verbal knowledge and specific modes of work coordination that come into play in the fabrication of material artefacts. Socio-technical system in contrast refers to the distinctive technological activity that stems from the linkage of techniques and material culture to the social coordination of labour. Thus, the proper and indispensable areas of social anthropology of technology include studies on techniques, socio-technical systems and material culture. Technology is described as a component of an ideological and cultural complex. It implies knowledge about the specific techniques, which comprise it. Technology in agriculture indicates the methods, systems, and devices, which are the result of knowledge used for agricultural practices.

Traditional technology involves the use of manual methods, tools and techniques used for the practical purpose. This form of technology is linked with the available local resources such as man and the surrounding environment; therefore it is easily available and accessible to the community of farmers. This technology is often simple and less energy consuming. The input cost is much less than in the modern-scientific technology. Sometimes anthropologists describe its practise in agriculture leading to a sustainable agricultural practice because it is based on the available local resources and mainly local knowledge. The crops adapt best with the regional climate and environment.

Farming in tribal areas is mainly based on the use of local knowledge and traditional technology. This traditional technology in tribal and rural areas is intertwined with that of traditional ways of life.

Modern technology, on the other hand, is highly complex and market friendly but less environmental friendly. The labour and time consumption are less in modern technology and it is favourable for better production, but energy consumption is more. It is expensive and risk pervasive. Accessibility to it is difficult in rural set-up due to the complex nature of modern technology. Skills and techniques to handle it need proper extension of services both from public and private domains. So the promotion of new technology, i.e. a modern-scientific technology, requires sufficient public support (Mann, 1982, Pachuri, 1984, Rao, 1988). However, modern technology is essential to certain extent to cope up with the rapid increase in population and growing demands for food. But the supply side should be stronger. There should be timely supply of inputs such as skilled labourers, capital, and research and extension facilities. Simultaneously, there should be other employment opportunities for the people followed by less labour absorptive nature of modern technology. Thus, without understanding the entire range of problems, it is difficult to introduce new technology; and without proper introduction it is difficult to think of sustainability.

Both traditional and modern-scientific knowledge and technology have their limitations and opportunities. Farmers try to use the new technology introduced by the formal organisations. Before taking up new initiative a farmer thinks of several factors such as socio-cultural, economic and ecological factors. The thinking process varies from individual to individual and community to community. Therefore, it is often essential to judge the farmers' rationality in adopting new technology apart from taking other factors into consideration (Gupta, 2005).

Agriculture in every society has undergone changes. Over the years, the changes in knowledge and technology adoption have been influencing the agricultural practices. Modernisation has been introduced with the change in knowledge, information levels, perceptions and concepts. This technological change according to Parrayil (1991) is the outcome of activities that humans engage in through their collective or individual organisational structures to optimise their resources, subject to constraints imposed by their own limitations in tandem with that of environment. Saviotti (1986) pointed out that any technological change is supposed to be having two characteristics which are complementary to each other. The interface between the internal environment of the technological system and the external environment is represented by the service characteristics, which provides the link between the technology and external environment. Foster (1962) described that the technological development as a particular kind of change in the structure of the society, in patterns of culture, and in individual behaviour. He describes what happens in "planned" or "directed" or "guided" change, to distinguish the process from the evolutionary or spontaneous or unplanned changes that constantly occur. In order to understand how these domains of change interrelate he spoke of "the three systems." First, a system of social relations, in which persons and groups are linked together by rights and duties, by expectations and obligations; second, a system of group conduct, which predisposes men to think and behave in normative ways according their perceptions and circumstances; and third, a system of individual cognition and behaviour, which underlies the first two systems, and rooted in biology and life experience, determines how the individual will react in a given situation. The three systems, the social, the cultural, and the psychological, are called for short as society, culture and psychology.

Technology Life Chain: The life span of any technology, agricultural or non-agricultural, consists of four distinct stages. All of which taken together, form the technology life cycle. As descried by Sharif (1986) the stages of the technology cycle are as follows:

- *Innovation Stage:* This represents the birth of a new product, material or process resulting from R & D activities. New ideas are generated by need pull and knowledge push concepts.
- *Syndication Stage:* This stage represents the demonstration and industrialisation of a new technology (product, material or process) with potential for immediate utilisation.
- *Diffusion Stage:* This represents the market penetration of a new technology through acceptance of the innovation by the potential users of the technology. Both supply and demand side factors jointly influence the rate of diffusion.
- *Substitution Stage:* The last stage represents the decline in the use and eventual extinction of technology due to replacement by another technology. Many technical as well as non-technical factors influence the rate of substitution.

Though his discussion didn't focus much on adoption of technology, but it is the stage which is often linked and overlaps with the diffusion stage. But in general, the adoption stage comes after diffusion stage in the technological cycle.

KNOWLEDGE, INNOVATION AND ADOPTION OF TECHNOLOGY IN AGRICULTURE

Local Knowledge and Farmers' Innovation

Local knowledge plays a crucial role in the innovation of technology. Local knowledge is the Indigenous Knowledge (IK), which is the systematic body of knowledge acquired by local people through the accumulation of experiences, informal experiments, and intimate understanding of the environment in a given culture (Rajasekaran, 1993). The IK is shared and communicated orally, by specific examples and through culture. Indigenous forms of communication and organization are vital to local level decision making process and to the preservation, development and spread of indigenous knowledge. This indigenous knowledge leads to indigenous innovation in the form of ideas, technology and practices. Broadly, the traditional technology system in agriculture is characterized by the limited interaction with outsiders and researchers of modern-scientific world. Indigenous knowledge in agriculture is also a similar system by which the farmers innovate in technology. The technology is shared and exchanged with others through communication. Knowledge of local innovations in agriculture includes crop breeding, water harvesting, soil management, conservation and processing. Indigenous agricultural innovations have continued to be important as most of the local grown food is for local consumption. Efficient organizing framework is most important for effective diffusion of knowledge and innovation (IK Notes, 2006). The IK Notes' study (2006) in agriculture in Africa mentions the promotion of the creation of local knowledge sharing networks to help innovators share their inventions with potential users and other innovators to both gain recognition for their work and to increase knowledge generation for further innovation. The study bases its observations on interviews with community-based innovators. It calls for public support for creating or fostering local knowledge sharing networks. The policy objective of a local knowledge-sharing network should be to find workable strategies to increase allocative efficiencies, increase their scale effects, and stabilize their growth in local economies. The policy would have to first stimulate a need to share knowledge among disparate innovators. Second, the policy would have to provide for knowledge "connections" to enable innovators, adopters and intermediaries to interact, for innovators to enhance the innovation process, for adopters to find solutions to their problems and for intermediaries to help connect and support interactions or improve the knowledge-sharing environment.

INNOVATION

Farmers' Innovation

Innovation is featuring the new ideas or methods (*Oxford Dictionary*, 2005) for practical purpose. Barnett (1953, 7) defined, innovation as any thought, behaviour, or thing that is new because it is qualitatively different from existing forms. Strictly speaking every innovation is an idea, or a constellation of ideas; but some innovations by their nature must remain in mental organisations only, whereas others may be given overt and tangible expression. Innovation is therefore a comprehensive term covering all kinds of mental constructs, whether they can be given sensible representation or not. Innovation is a part of every community. In agriculture both farmers and researchers are innovators, but not all. First of all the realization that farmers innovate stems from individual contacts, e.g. a project employee or a field researcher gets to know an innovative farmer and then documents his/her activities. This evidence is by its very nature anecdotal when the contact is not a result of random sampling. Secondly, a resistance to acknowledging farmers' innovative capacity has meant that it has hitherto been more important to show that farmers innovate, than to collect statistically sound data (Waters-Bayer *et al.*, 2001). Farmers are the silent and natural innovators. Their technology is developed through frequent experiments and practical experiences (Nandy, 1987; Gupta, 1995). Some farmers are innovative and their contributions are noteworthy. These innovations transfer from farmers to farmers through communication, which is very significant. As described by Bellon (2000), farmers cannot and do not wait for the scientists "go ahead" signal to disseminate approved and validated innovations or technology. New ideas spread primarily through farm visits and other forms of farmer-to-farmer communication. This approach depicts that the farmers are both innovators and adopters. Invention is an inherent part of all farming systems and have historically accounted for much technological change in agriculture (Ridgley and Brush, 1992).

Others' Innovation

Apart from farmers, the researchers, public and private workers, and administrators, industries also innovate in technologies or policies in agriculture. The innovation and invention of researchers do correspond with the effort put in the laboratory work than in the real field. Researchers develop technology for a larger population. Their experiment is drawn from a larger sample. The focus on one community and to a particular climate is less. Therefore, it may succeed in one climate and become failure in other. Again, the modern scientific technology is expensive and risk pervasive. These may not be suitable to a larger mass of marginal communities. However, the importance of scientific development is still crucial because of its high productive efficiencies.

The technology developed by both the groups is important. Traditional technology is the base to develop scientific technology. Scientists need to understand the local practice, knowledge and innovation while developing a technology package for the local people. A farmer does not need scientific technology in each and every practice. However, where the farmers' knowledge doesn't give expected result despite continuous effort they invite support from others. The failure of local knowledge seeks the support of scientific knowledge. But, all the technologies developed by the scientists are not suitable to all the situations and environment. Farmers have to manage in that situation with their knowledge and innovation. They become failure if their idea goes wrong. Therefore, effective dissemination of selective technology needs to be introduced.

Modernisation and Innovation

According to Hagen (1964), creativity is the prerequisite for modernisation. Creativity comes from innovative personality. Tradition itself is a product of long period of growth and change. Tradition is never totally replaced. It is not irrational in total. As far as agriculture is concerned, there are opportunities in both traditional and modern agricultural practices. Both have rationales. With regard to innovation, modernisation in agriculture is also an innovation and development from traditional agriculture. Thus, there is close proximity between tradition and modernity.

Adoption of Innovation

In the 1940s, two sociologists, Bryce Ryan and Neal Gross "published their seminal study of the diffusion of hybrid seed among Iowa farmers" renewing interest in the diffusion of innovation S-curve. The now infamous hybrid-corn study resulted in a renewed wave of research. "The rate of adoption of the agricultural innovation followed an S-shaped normal curve when plotted on a cumulative basis over time". This rate of adoption curve was similar to the S-shaped diffusion curve graphed by 'Tarde' forty years earlier.

Ryan and Gross classified the segments of Iowa farmers in relation to the amount of time it took them to adopt the innovation, in this case, the hybrid corn seed. The five segments of farmers who adopted the hybrid corn seed, or adopter categories are:

1. Innovators;
2. Early adopters;
3. Early majority;
4. Late majority; and
5. Laggards (Fig. 1.1).

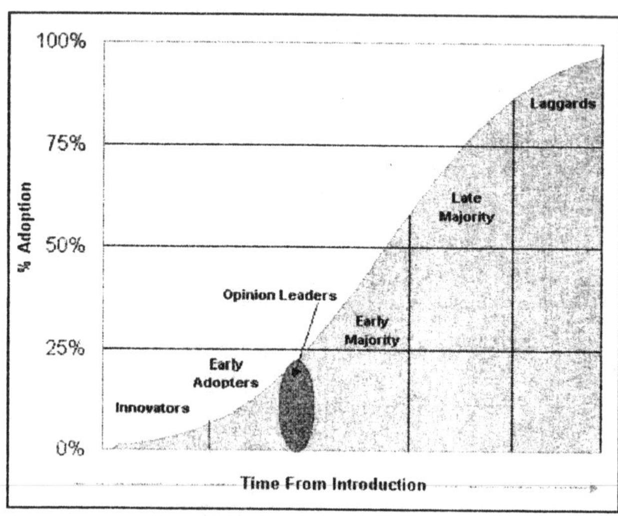

Fig.1.1: **Segments of farmers and time taken for adoption of innovation**

CROP SELECTION AND PRODUCTION

The selection of crop and practice of cultivation are guided by several factors. Among the important factors, the farmers' observation and frequent experiment with the particular crop is important. Farmers select some crops which are suitable to a particular agro-climatic condition, market arrangement, and other institutional arrangements. As described by researchers supporting participatory approaches, the farmers take many situations into account for taking up a new practice that may be selection of seeds, use of agricultural tools and implements, and others. It is not that the population pressure alone plays important role, but there are several other factors such as farmers rationality, interest, and attitude like psychological factors; economic factors like farmers' capacity of accessing human and capital resources etc, and environmental factors like climate, soil quality, rainfall, etc play important role. The cultural factors like beliefs, norms and values of the community are also important for the above purpose.

III

FACTORS AFFECTING ADOPTION OF TECHNOLOGY IN AGRICULTURE

Multiple factors such as demography, climatic and environmental factors, institutional and organizational arrangements, etc, are crucial in the knowledge production, innovation and adoption in agriculture. Okunado (2006) found some of the important factors that affect adoption. These are broadly classified into six parts such as characteristics of innovation, characteristics of adoptions, cultural factors, change agents, government policies, and environmental factors.

SOCIO-ECONOMIC FACTORS AND ADOPTION OF TECHNOLOGY

Socio-economic factors have been the important factor for the adoption of technology in agriculture. Though there are a large number of studies available in this regard, only a few are discussed here.

Sachchidananda (1972), in his study of Shahabada district of west Bihar, focused on various socio-economic factors related to agricultural innovation. He also described the various issues involved in agricultural sector in tribal areas like caste factor, lack of communication, personal and psychological factors.

Singh (1982) has rightly pointed out that the issues of land alienation, depeasantisation, and bondage are common in tribal and rural societies. His observation is that in many cases the alienation of land was not only between tribals and non-tribals but also among the tribals themselves. Land alienation is one of the factors to affect to adoption of new technology in agriculture in rural areas.

Das (1982) in his study of Zubza, an Angami tribal village of Nagaland discussed the problem of land alienation. Most of the Angamis sell their land to non-Angamis. This large transfer of land has affected the Angamis for further innovation and adoption of alternative technology.

Kulkarni (1974) has cited several problems persisting among tribal peasants. He mentioned that the predominant issues among them are the problems of land alienation, exploitation by moneylenders, and adherence to primitive form of technology, etc.

Barlett's (1980) and Stanford's (1994) studies pointed out that organisational aspects are crucial to decision on agricultural production. Kinship structure, clan groups, gender groups, labour organization, self-help-group (SHG) formation, etc play crucial role in adoption of innovations.

Other than the above factors, some other social factors like lack of educational support, health, infrastructure, economic factors like lack of investment, market supports etc affect the innovation and adoption process in agriculture.

CULTURAL FACTORS

Agricultural cycle in many tribal and rural areas is associated with ritual cycle. Most tribes perform rituals before and after agricultural production. They celebrate festivals during and after harvesting. Cultural traits like ethos, norms, and values are often responsible for the adoption and innovation of technology. Gonds in Andhra Pradesh and Kandhas in Orissa have different festivals and rituals associated with agricultural production. All these practices are associated with the traditional agriculture.

Mahapatra (1982) from his study on Santal in Mayurbhanj district of Orissa noticed the strong association of culture and technology. He discussed the role of rituals and their linkage with the agricultural production. The Santal celebrate number of festivals and rituals that are associated with the agricultural practices. The adoption of the technology is thus shaped by these traditions and customs. The other problems associated with the adoption process are smallholding size, lack of irrigation and the incidence of indebtedness. All technological changes and innovations, therefore, presuppose specific cultural determinants.

Kattakayam's (1983) study among Uralies in Kerala depicts that ritual performance is must in different stages of agricultural practices. They seek trouble free agricultural production with suitable climate, free from wild birds and animals, no failure of crops, etc. Das (year not mentioned) pointed out the significance of social factors that are concerned with belief system in productive mechanism of Bhumias of Koraput district, Orissa. He also explained the harmony that is maintained between traditional belief system and new ideas related to production through the rituals by appeasing various spirits and deities. He has justified the significance of traditional knowledge system (TKS) in agricultural production. These associations of rituals, festivals with the agricultural practice influence the adoption of traditional agricultural practices.

Villarreal (2000) pointed out that land ownership, access to other productive system and the organization of agricultural production and technological adoptions are influenced by cultural practices and traditions. For example, rules of land inheritance (by lineage, gender and/or other culturally determined characteristics) are core determinants of effective access to land. Cultural aspects are thus of central importance for the understanding of appropriate inventions in agriculture.

Mead (1954) focused relation between cultural patterns and technical change. As discussed by her, the new knowledge can be put to use only as the old behaviours, beliefs, and attitudes are unlearned and the appropriate new behaviours, beliefs and attitudes are learned. She further pointed out that both individuals and groups are important as far as cultural change is concerned. In all technical change, the individual person is both the recipient of change and the mediator or agent of change. Therefore, the individual's integrity as a person, his stability as a personality, must be kept ever in focus as the living concern of all purposive change.

KNOWLEDGE, INNOVATION AND RATIONALITY

Fujisaka's (2000) study on Philippines Tupi farmers' rice production strategy, their knowledge and technology adoption shows that the farmers select rice varieties based on yield, eating quality, durable disease resistance, duration of production, and weed competitiveness. Tupi rice and maize farmers plant three crops a year while facing the problems of crop diseases, weeds and uncertain rainfall. They plant maize of modern varieties in low land and traditional variety rice in the upland in the wet season. They combine tractors and animals for initial land preparation, and use a furrow secondary tillage technique.

Rajasekharan & Warren (1995) discuss the role of local taxonomies of rice varieties in the decision-making process leading to farmers' selection of rice varieties in south India. Choosing a rice variety depends on number of decision-making factors. In order to effectively implement the program, cognitive strategies and organized body of knowledge are essential.

Gupta (1995) while drawing upon peasant's innovation in Bangladesh found that farmers have excellent innovation in storing and keeping the plants. Similarly farmers also have knowledge and innovation in protecting crops from insects and pests.

ROLE OF ATTITUDES IN ADOPTION OF TECHNOLOGY

Attitude is one of the major factors in the adoption of technology in agriculture. The farmers having right attitude to development accept the things, experiment and frequently try to search for the positive result.

Personal behaviour or group behaviour affects the traditional practices or adoption of new practices in agriculture. Besides the above personal or psychological factors, cooperative activities, personal effectiveness, political participation, etc are the important socio-psychological dimensions of agriculture of development. (Sachchidananda, 1972; Haimendorf, 1979).

FARMERS' RATIONALITY

The innovation and the adoption of new technology depend on several cultural factors including the rationality. Rationality implies that when a farmer faces with a set of alternative actions, he will evaluate and rank them according to his own particular preferences. He will choose such course of action, which he ranks the highest, in other words he will try not to forgo a high preference action. It is just as important to consider what he would have to forgo as to consider what he will receive, since preference may be ranked into pairs or sets (Ortiz, 1970).

Farmers adjust their decisions to the range of expectations and not to any one of the more possible outcomes. When we talk about rationality it is significant to consider outcomes in terms of utility than in terms of quantity. Ortiz (ibid.) also pointed out that rationality doesn't imply that tradition; habit and even impulse do not play a role. Traditional means of cultivating can play important role in rational behaviour. The adoption of innovation depends mainly on the farmers' rationality. The people who adopt think several times before going for practice.

ENVIRONMENTAL FACTORS

There are studies like Adejuwon's (1962) study of the intensity of cocoa production in western Nigeria, Beals (1974) study in South India, Perin and Winkelmann (1976) studies of corn and wheat production and many others' studies indicated that environmental factors like altitude, soil type, rain fall, temperature, wind, agro-climatic zone and topography, etc play important role in the production process. Towards the last half of the twentieth century leading anthropologists developed the separate discipline of anthropology called agricultural anthropology, where Netting (1974, 1992) was the most prominent figure. He emphasized how the agricultural systems affect the environment and environmental feedback in response to their agricultural systems, and these perceptions affect values, knowledge and behaviour. Epstein (1962) also notes the role of occupational diversification. She writes that in the unirrigated village of Dalena, economic diversification diluted caste lines, whereas in irrigated Wangala village, farmers prospered as they changed their mode of cultivation. But their traditional social system, based on hereditary obligations in intra-caste and inter-caste relations remained unchanged.

ROLE OF PUBLIC AND PRIVATE INSTITUTIONS FOR DEVELOPMENT OF AGRICULTURE

Many of the views that other than the above factors, the public and private interventions and the influence of the macro-micro policies matter a lot for the technological adoption. The emergence of spatial pattern of crops with communal bias requires a sub-regional organisation of agriculture with infrastructure-markets, transport services and systems, extension and information services by the help of public private partnership. The property structure and land relations are the major deciding factors for change in technological adoption. Institutional actors such as private agencies, NGOs, universities and research institutes, research foundations, farmers' organizations, state government, and local governments at the regional and sub-regional levels play crucial role in bringing innovation. Some of them are directly involved and some are indirectly, but both are crucial in shaping the changing process. Agricultural knowledge and information flow through a variety of channels, including private sectors, universities and commercial suppliers of agricultural inputs and equipments.

NEW ECONOMIC POLICIES

The new economic policies adopted after 1992 with the framework of WTO policies have intensified the rural problems. The new policies in import and export of agricultural products, decontrol of fertilizers, etc have affected the poor farmers and small peasants. The problem of free trade policies and the liberalization of imports of certain agricultural commodities like raw cotton; edible oil and pulses have strongly affected the small farmers. With the decontrol of fertilizers like phosphatic and potassic fertilizers in 1992, the central government in India is not able to provide direct subsidy to farmers that lead to negative consequence. Similarly, state governments due to financial constraints eliminated subsidies in many inputs like electricity, water and fertilizer, which do not make any sense for the poor farmers (Acharya, 1997). While looking at the rising issues in post economic reforms period Rao (2001) mentioned that due to new economic policies many significant values of indigenous knowledge system (IKS) are gradually losing ground with the advancement of time. Ramanjeneyulu (2006) noticed the change that is taking place in the wake of new economic policies. He pointed out that the technology development in present context has become market oriented rather than need based. With the change in trade policies, the production and produce have become market oriented. Only those technologies, which can be effectively marketed, are developed by the researchers in both public and private sectors. This makes the resource poor farmers depend more and more on markets for their inputs and external markets for their outputs; and this is how they are deceived at both the ends. As the market dependency has increased agriculture has become more capital intensive. To cover up the costs and to meet their other needs the farmers have started growing only those crops and varieties, which market demands. Thus farmers have lost control both over inputs and outputs.

The new government policies, which are indirectly or directly meant for agriculture, do not favor a large mass of marginal farmers belonging to tribal and rural communities.

Government's policies in the form of privatization of electricity sectors and fertilizer sector, imposition of heavy taxes and reduction in subsidies are directly responsible for the imbalances in technological adoptions and production processes. A large gap between the needs of the farmers and the public and private extension is noticed in agriculture.

OTHER FACTORS

Farmers adopt or do not adopt a technology or select or reject a crop depends on the economic factors, environmental factors, awareness and persuasion, and other risk factors, etc. If a crop is not adaptable to a climatic situation or weather condition and geographical location, the chance for adoption of that crop is less. If inputs required for agriculture are expensive for the farmers, the supply of inputs becomes irregular, the infrastructure is poor, etc, then the adoption of certain technology also becomes difficult.

Small farmers and marginal landholders do not easily switch from field crops to horticultural or other high value crops as these are much more labour and other input intensive and require more post-harvest handling. To adopt a new crop or technology, farmers need more than just technology. Risk aversion is identified as a significant impediment to what would seem to be a rational diversification on the basis of average profitability of alternative crops which, in turn, is affected by the attitudes of individual farmers and the nature of technology (Singh, 2005).

Technology is the part of the knowledge system of the community itself and has a content and process which has persisted through the generations. It is learnt by the individuals through a process of observation of the elders at work and imbibing the folklore and accumulated wisdom of the community. In the course of this transmission of knowledge, it has been enriched and integrated into the collective consciousness.

Access to technology differs between the non tribal and tribal communities leaving in mainstream of the societies and the isolated tribal communities. For the non tribal world at large, the range of operation of technology encompasses the totality of global endowment of resources, where as for the isolated tribal populations the choice is considerably restricted. Pawar (1972) reported that majority of the tribal farmers in India have not yet tried improved variety of seeds on their farms. Application of manures and fertilisers to the crop is also limited. He could substantiate it from his fieldwork among tribals in northern Gujarat. The case of the tribal communities in Orissa have similar situation, the access to modern technology is less among tribals. Mahapatra (1978) made a few points about the resisting factors for accessing new technology by Santal tribe in Mayurbhanj district in Orissa. He pointed out that the diffusion of knowledge of improved agriculture is limited by the communication gap in tribal society. Firstly very few *gram sevaks* and agricultural extension officers know the tribal language and can carry conviction, the incentives to pick up the new technology, and the new process is generally absent due to lack of feasibility of application. Small landholdings, lack of irrigation, and indebtedness are some of the inhibiting factors. Thakur (1997) also viewed the communication barrier to technology. The some other important issues that he focused are lack of women oriented training and agricultural extension workers and irrigation problem. Modern agricultural technology is at a very high level and is too sophisticated to be suddenly adopted by the tribals. The adoption of new techniques in farming requires the communication of information about them. When viewed as a process the adoption of a new technique requires several different kinds of information in agriculture. But the tribal areas are as such lagging behind to get update information in agriculture.

With regard to technology development and transfer in tribal areas, some of the important issues are highlighted again. The technologies that are developed outside cannot be accepted suddenly by the tribal communities. The reasons are social, cultural, economic, environmental, etc. In present situation, the tribal communities in India are more or less aware about new technology and new practices. New technology is being accepted by the tribals even based on the surrounding factors.

IV

THEORETICAL APPROACH

The study on emerging trends in agrarian transformation has broader dimensions and fits into several disciplines. It is much more than a mere change in agriculture. For understanding the essence of such a complex process, no single theoretical approach will be adequate. Hence insights derived from relevant approaches have been chosen for the understanding of agrarian transformation. It is already discussed under conceptual understanding that the study on agrarian transformation occurs in the due course of time. It also discusses how in the process of technological transformation, the institutional intervention and policy factors play significant role.

In view of above, the theoretical approach has been considered under three major dimensions such as technological dimensions that includes development, dissemination, and adoption of technology, institutional aspects, and policy aspects.

DEVELOPMENT OF TECHNOLOGY

There are several approaches developed by the anthropologists, sociologists and economists to find out the reasons for the development of technology. Some of the approaches directly and some are indirectly linked with the discussion of technological development.

DEMOGRAPHY AND POPULATION

Boserup (1965) hypothesized that agricultural intensification, defined as increasing the annual returns from land, is an adaptation to the need to produce more food for growing population living on a fixed land area. Even in lowland tropical environment, where the climate does not favour better production by the adoption of new technology, the carrying capacity of the land is increased. She has challenged the traditional idea of the relationship between technical innovation and population growth, i.e. population growth is a function of technological innovation. PG=f (TI), which hypothesized that population growth is derived from new opportunities to increase reproduction following upon technical innovation (Malthusian approach). But her proposition is that technology innovation is stimulated by population growth (Schneider, 1975).

In a neo-Malthusian model, climate an independent variable has a positive effect on the initial dependent variable, agricultural production: when the climate is favourable, agriculture production is high; when it is unfavourable, production is low. Agricultural production has a positive effect on a second dependent variable, population density: when production is high, population density is high; when production is low, population density is low. Thus climate has an indirect effect on population density through its effect on technology development and agricultural production (Hart, 1990).

SUBSTANTIVISM AND FORMALISM APPROACH

Polanyi (1944) provided two meanings of economy. These are substantive and formal. Substantive meaning explains the man's dependence exclusively on nature and its fellows. The integration of natural and social environment broadly defines the substantivism, which results in supplying the man with the means of material wants to give satisfaction. This indicates that primitive agricultural system is more substantivistic, which never implies the choice or insufficiency of means. Whereas, the formal meaning of 'economic' derives from the logical character of the means-ends relationship as apparent in such words as economical or economizing. It refers to a definite situation of choice, namely that between the different uses of means induced by an insufficiency of those means. If we call the rules governing choice of means as the logic of rational action, then we may denote this variant of logic with an improvised term a 'formal economics'.

According to Le Clair (1968) the formal economic process can be seen as a series of four activities such as:

1. *Supply and demand:* the value of goods and services will increase when they become scarce or when demand increases.
2. *Economization:* people will economize or allocate scarce resources among alternative ends.
3. *Maximization:* when engaging in production and exchange activities people will seek to get the maximum possible returns or benefits.
4. *Rationalization:* people will seek the means to improve their productive or exchange capacities.

From the above approach it is derived that the development and adoption of technology in subsistence economy is limited. This is because of no choice and no insufficiency of means. The economy is nothing but guided by nature and the social relationship. This is the main characteristic of subsistence economy. But when people move from subsistence economy to commercial economy, the logic of formalists' economics in terms of the four above mentioned principles operates.

DISSEMINATION AND DIFFUSION OF TECHNOLOGY

Technology development occurs at various levels as an effect of various factors. But the important discussion is how these innovations are transferred to farmers. It is known that the farmers exchange the technology through communication. That is easy to say, but how these exchanges occur? Many a studies have tried to explain how the technology is transferred to farmers. A few of them are as follows.

THE DIFFUSION OF INNOVATIONS AND ADOPTION

In his book *Diffusion of Innovations,* Rogers defines the diffusion process as one "which is the spread of a new idea from its source of invention or creation to its ultimate users or adopters". Rogers defines "the adoption process as the mental process through which an individual passes from first hearing about an innovation to final adoption".

FIVE STAGES OF ADOPTION

Rogers breaks the adoption process down into five stages. Although, more or fewer stages may exist, Rogers says, "At the present time there seems to be five main stages". The five stages are:

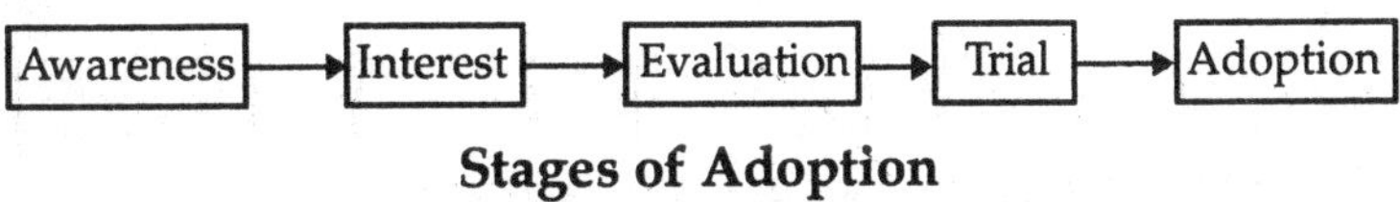

Stages of Adoption

Source: Rogers, Everett M. (1962). *Diffusion of Innovation,* The Free Press. New York.

In the awareness stage "the individual is exposed to the innovation but lacks complete information about it". At the interest or information stage "the individual becomes interested in the new idea and seeks additional information about it". At the evaluation stage the "individual mentally applies the innovation to his present and anticipated future situation, and then decides whether or not to try it". During the trial stage "the individual makes full use of the innovation". At the adoption stage "the individual decides to continue the full use of the innovation" (Fig. 1.1).

INSTITUTIONAL FRAMEWORK

Institutional framework is very important in implementing new strategies and promoting new agricultural practices. So after green revolution the development of institutional structure has become crucial in order to effectively transmit the new technology. There is change in adoption of knowledge, information and technology after the new institutional arrangement, which is broadly guided by the regional and state policies for agricultural development. Institutional influences lead to significant changes. Development of both public and private agencies is important for supplying new inputs such as knowledge, information, technology, credit, and marketing supports.

Institutional framework stimulates the production mechanisms by successful introduction of technology. Institutional framework is important because organizational structure is important aspect of developing the production system. The behavior of the staff and the researchers who are usually responsible for dissemination, intervention and dissemination of technology are important for continuity and longevity of the adoption. Despite the importance of institutional arrangement in changing technology adoption, the anthropologists working in agricultural development have given less emphasis to this model.

Extension of services is a very crucial factor for agricultural growth. Extension of services is done by both public and private agencies. In rural and tribal areas where the provision of government services is very weak, NGOs may be the main providers of extension of services. Not only do they provide the services but they are also responsible for developing many of the methodologies for research and extension work, which are subsequently adopted by the public sector (Farrington and Amanor, 1991).

The institutional arrangement not only focuses on the behavior of the members in the institutions and organizations but also on the approach of dissemination of technology, which has been very important in agriculture, particularly, for the effective use of technology. The institutional perspective thus views organizational design not as a rational process but rather one of both external and internal pressures which lead organizations in a field to resemble each other overtime.

EXTENSION APPROACH AND TECHNOLOGY TRANSFER IN AGRICULTURE

Over the years in Indian agricultural extension programs, the transfer of technology (TOT) approach has been given importance. The TOT approach is a top-down approach, where the local farmers are least consulted. The incorporation of indigenous knowledge in transfer of technology is very less. The technology are usually developed in the researchers' lab and transferred directly to the farmers' field without mutual consensus. The same system of approach has been followed in both urban and rural areas and non-tribal and tribal areas. Many studies by Bellon (1995), Warren (1995), Chambers (1997), and others have criticised the top-down approach because it takes a unilateral position in the technology dissemination.

In the history of evolution of extension approaches, the Farming System Research (FSR) approach took a disciplinary position in technological dissemination. FSR does not reject the local knowledge and innovation. In FSR approach the farmers' participation is crucial in defining their own problems and sharing the knowledge with the researchers. Willigen (1986) suggested Farming System Research (FSR) could give better understanding of the culture, environment and economic conditions of the community. Many of the researchers like Erwin, (2000); Shaner, *et al*, (1982); Conway and Barbier, (1990); and others suggested the similar approach, i.e., farming system research approach that was developed by Willigen in 1980s. However, Participatory research is more accepted these days seeing the importance of people's participation on agricultural development from both on-farm and off-farm sides.

PARTICIPATORY RESEARCH APPROACH AND TECHNOLOGY DISSEMINATION IN AGRICULTURAL DEVELOPMENT

Though many approaches have been developed for studying agricultural production such as technology intervention or mechanization approach by social and biological scientists, farming system research is selected based on its appropriateness in understanding agricultural practices and local knowledge in a particular community. After failures of the top down approaches in 1950s and 1960, an effort has been made to create developmental programs and services to fit local farmers in decision-making. Attention has been paid to local values, knowledge and social organization although the idea is still far from universal. New research approaches have been variously labelled as farming system research, participatory research, and agro-ecosystem analysis (Shaner *et al*, 1982; Conway and Barbier, 1990; Willigen, 1986). There is an understanding that farming requires holistic analysis. This needs to be done by interpreting constraints and opportunities related to local environmental, cultural and historical factors, in the choice of crops, livestock and farming procedures, and agricultural policies on local forms. Essential to these systems is the role of community, family and domestic cycles in agricultural decision making. The goal of farming system research (FSR) is to first define and account for the local systems, its subsystems and variables. It then becomes a matter of carefully matching highly variable technical solutions (including the decision not to intervene) to a regional agro ecosystem (Ervin, 2000). Farming participatory research (FPR) was widely accepted followed through the cooperation between researchers and the local peasants, farmers and the off farm experts. Participatory techniques or Participatory Rural Appraisal (PRA) techniques are some of the most important techniques for data collection. Participatory research techniques will be helpful in following stages. These are diagnostic or description stage and design stage. The role of anthropologist in diagnostic stage is more important. Direct observation will be the part of it.

ADOPTION OF TECHNOLOGY

The entire process of technology adoption occurs in slow and steady process through various stages. Number of factors and variables are responsible for the adoption process, but among which the role of public and private sectors, their policies and approaches are crucial and convincing factors in order to cope up with the various socio-economic and ecological factors. The combination of local knowledge and technology with the scientific ones or vice-versa has been significant for a sustainable approach. Definitely all the factors influence one another for the innovation and adoption of technology. The technological innovation takes place both at the farmers and the researchers' level. Both innovate-based on their knowledge and understanding. The farmers consider a lot of factors which include knowledge, rationality, risk, external pressure, internal pressure, etc. while adopting the innovations.

Mead's (1954) observation is that the immediate need for agricultural changes arises from the great increase in population in recent years. In her book on "Cultural Patterns and Technical Change" she discussed some of the following dimensions:

Culture and Adoption of Technology

Mead pointed out that in many parts of the world the cultures adhere to the belief that man has no casual effect upon his own future or the future of the land; God, not man, can improve man's lot. This view exists in different forms. With examples from different communities in the world, she found that it is difficult to persuade such people to use fertilisers, or to save the best seed for planting since man is responsible only for the performance, and the divine for the success of the act. So culture and adoption of technology are interlinked.

Need for the Assured Result

While focusing need for the assured result she has illustrated how ineffective and disastrous it is without understanding local conditions. In Burma, for instance, deep ploughing introduced by European agricultural experts broke up the hard pan that held the water in rice fields.

The Valued

The programmes for improvement are sometimes rendered ineffective because of failure to take account of what the people specifically value. Values may range in intensity from strong preferences to highly emotional charged attitudes of a religious significance. With evidences of case studies, she addressed the need to understanding values, rationales, beliefs and attitudes of the communities for effective introduction of a new technology or implementation of a programme.

Understanding Culture for Technological Change

Introduction of new mechanisms, which does not meet the local conditions in the form of contour ploughing caused negative consequence in different regions, as in the case of introduction of cash crop, which has resulted in lowered nutrition. In this form, she delivered the idea of understanding culture required for the introduction of technology towards change.

Technology and Social Change

Mead focused, when the change is from subsistence agriculture to cash crops, a number of disruptive factors enter the picture. She pointed out that we now have a radical change from, 'making a living to earning a living' (1954, 181). According to her even when we only introduce new techniques we make a break with the sustaining tradition which gives security.

Foster (1962) has discussed several factors and the barriers in social change. They are broadly divided into social, cultural and psychological barriers. The farmers do not adopt something suddenly. The adoption takes place through frequent observation and experiments. Particularly, in rural areas the change in certain things is only possible after a long process. The change in customary behaviour and norms is very difficult. As the technological adoption deals with the change in other process, it is necessary to take various factors into consideration.

In rural areas, the involvement of public and private sectors is essential for their successful implementation of new strategy. The regional and macro-economic policies have been crucial in the changing practice. The changes that are occurring in agriculture in present scenario are thus guided by multiple factors. Changing cropping pattern and technological adoption are not only influenced by changing internal arrangements but also by the external change agents that include public and private agencies. Changes in agriculture are guided by the policy factors of which economic policy is a very significant one.

The framework thus represented below explains that the innovation at formal institutional level takes place based on local knowledge base and scientific knowledge base. Both researchers and government and non-government organizations innovate. These innovations are diffused and disseminated to farmers through extension and exchanges. Farmers who are the ultimate recipients take the decision whether or not to adopt the innovations. The farmers' adoption depends on multiple variables including knowledge, rationality, risk, external pressure, internal pressure, etc. If farmers could overcome the resistances from the above factors or some of the factors became suitable, then adoption of certain practice becomes possible. This ultimately leads to socio-economic changes.

There are also the roles of multiple factors such as socio-economic, ecological, political, etc. for the changing agricultural practices. New economic policies have certainly some effect towards this process. The significant approach on which the analysis based is the economic approach. However, this economic approach is supplemented by multiple approaches. Therefore, the model developed for this purpose is a combination of several approaches.

V

STATEMENT OF THE PROBLEM

From the review of literature it is clear that though quite large numbers of researches have been carried out on technology adoption in agriculture, not much effort have been made to understand the farmers' knowledge and rationality in choosing crops and selection of technology. There is gradual transformation from traditional agricultural pattern to modern agricultural situation in rural and tribal areas which needs specific approach to understand the transformation process minutely. Many studies have discussed the role of institutional factor in agricultural production. But the deeper understanding about the role of various agencies and their approaches in agriculture development, other institutional approaches and intervention mechanism in technology dissemination is yet less recognized. In most of the cases it is observed that the real issues in agriculture are mainly derived from the researchers' point of view than deriving it from the farmers' views. So far, we are not able to explain the real issues in agriculture in its transformation. Therefore, the real challenge is to find out the hidden attributes and explore various other abstract factors in understanding about the emerging trends in agrarian transformation. Agrarian transformation needs elaborate discussion keeping the emphasis on change in cropping pattern, technology adoption, institutional intervention and policy intervention rather than focusing agrarian relation and other aspects in isolation.

In the wake of new economic policies, many changes are observed in agriculture both in tribal and non-tribal areas. As this is relatively a new phenomenon, there are very few studies available for understanding the problems of people in the wake of new economic polices in tribal agriculture.

In view of above, here is an effort to cover various socio-cultural and economic aspects, institutional intervention and policy dimension for understanding about the agrarian transformation.

Keeping the above into consideration, the research is broadly focused on the following questions: What are the traditional agriculture practices and what changes are observed? How are the personal factors and the institutional factors responsible for technological adoption? How do the public and private agencies intervene in dissemination of technology in rural areas? Finally, how the agricultural system is affected by the new economic polices? These questions are focused by framing the following specific objectives.

OBJECTIVES OF THE STUDY

1. To understand local beliefs, perceptions, knowledge, and technological adoption in traditional agriculture.
2. To examine the trends in agricultural practices and technological adoption.

3. To assess the impact and effectiveness of involvement of public and private agencies in agriculture, and
4. To delineate the nature of transformation that is taking place in the wake of new economic policies and their implications for agriculture.

THE STUDY AREA AND THE METHODS ADOPTED

Kandhamal district is one of the main tribal populated districts in Orissa. The population of the Kandha tribe is more than 51% in the district. Next to Kandhas, the Panas dominate as majority among the non-tribal population in the district. The Panas belong to the scheduled caste. Their population is a little more than 18% of the district as per 2001 population.

The Kandhas are well known for their age-old cash crop production such as turmeric and ginger. Both shifting cultivation and settled cultivation have been important part of the Kandhas' economic activities. The production of such crops reflects their beliefs, perception, knowledge of production, and technology adoption. The Kandhas developed various indigenous mechanisms to adapt in the specific agro-climatic zone. The Kandhas have improved their vegetable cultivation practice over the years. Sericulture has also been taken up in many parts of tribal villages. It shows that the commercial cash crop cultivation is gradually shaping up in the region along with the change in technology adoption.

Panas are mainly considered as middlemen in agricultural trade. But agriculture too is the occupation of many Panas. In the contemporary situation, the Panas are also involved in cash crop production such as turmeric and ginger and vegetables.

In recent years, both the communities have adopted cultivation of different crops with different techniques and technological know-hows. With the involvement of agencies, both private and public, and implementation of developmental policies, agricultural practices witnessed further changes.

It is not only the agricultural occupation of the communities or the changes in technology adoption, but other factors like significant association of culture and agriculture as depicted in earlier studies led to this study of rural communities in Kandhamal district. Moreover, the early literature depicts a lot on the traditional behaviour of the Kandha community, which is reflected in the form of beliefs, knowledge and behaviour. The Kandhas and the Panas are the inhabitants of the Eastern Ghats. A few years back the district was declared as the best agro expert zone for spices (APEDA, 2002).

Technological breakthrough is a new vision in the community. After green revolution many changes have taken place, particularly in the field of technology adoption and crop production. Involvement of research institutions and cooperatives is a significant change in the villages under study. It is essential to discuss the role of multiple factors such as socio-economic, ecological, institutional, etc and their role in technological changes in the villages.

Based on all these factors the district was thought to be suitable for pursuing the objectives of the study.

PILOT SURVEY

In order to have preliminary and basic idea about Kandhamal district, a pilot survey was conducted in May-June 2005. During the pilot survey the selection of the villages was done after proper scrutiny. Preliminary information about the district and agricultural situation was collected from the officials of various institutions that include District Rural Development Agency (DRDA), District Agricultural Office (DAO), Statistics Department, etc. After having preliminary information about the district, G. Udayagiri block was selected, which could provide the base for pursuing the objectives of the study. This block represented the presence of both social groups, Pana and Kandha, who are predominant in the district. Vegetable cultivation has taken an important place in the block. Irrigation is moderate. There is instance of both traditional and modern agricultural practices (DAO, Phulbani, 2005). Krishi Vigyan Kendra, Additional District Agricultural Office, and Regional Research Technology Transfer Station are some of the important agricultural research centres in the block. The presence of NGOs and other voluntary agencies like Samanwita, KASAM, Swati, etc in the block was also important for selecting the block.

After collecting both primary and secondary data from the regional research centres, block head quarters, and some people of the locality and others, it could be known that the Katingia panchayat of G.Udayagiri block is known for vegetable cultivation. They mainly grow French beans and turmeric, and they are well known beans cultivators. In addition, it was also found that after this Katingia panchayat, the people of other regions have started growing beans and some other vegetables crops. This information about the panchayat was one of the reasons to select the panchayat for undertaking the study.

For the purpose of understanding agrarian transformation in tribal areas, two major inhabitant groups in the block i.e., the Kandhas and the Panas were taken into account. The two groups are closely associated both from social and economic fronts. They are the major inhabitants in the district. The selection of the villages was done considering the presence of these two groups. The villages also represent the inclusion of both Christians and Hindus. Moreover, the villages being centrally located in the district has medium access to market and other institutional supports from both public and private agencies. Two villages such as Sudhipara and Laburi were selected for the intensive fieldwork as they are of manageable size and suitable for the specific objectives of the study.

The main fieldwork was conducted from August 2005 to May 2006 in the selected Katingia Panchayat of the G. Udayagiri block.

RAPPORT ESTABLISHMENT

It would have been difficult to collect data if I had not interacted with the village leaders at my first visit. The rapport establishment with the village people through the village leaders made the work easy to be conducted. Within no time the rapport was established with some of the villagers and a house was given to me for staying in the villages during the entire fieldwork. The rapport was so strong that, there was no barrier in collecting data while discussing formally and informally from the respondents from any place.

DATA COLLECTION

Data collection was from both primary and secondary sources. The primary data collection was by the use of qualitative research techniques such as observation, interview (structured and non-structured one), case studies, schedules and focused group discussion. Participatory Rural Appraisal (PRA) was also a meaningful approach for the data collection. This was primarily meant to cross check the data collected by the traditional research techniques.

OBSERVATION

Observation was one of the most important tools used for collection of data with regard to rituals and festivals associated with agriculture and agricultural practices; processes in ploughing, land preparation for different crops production; understanding selection of seed varieties, sowing, use of tools and techniques, application of modern technology, processes of cultivation, irrigation pattern, distribution and sharing of water for irrigation, organisation of farmers' society and other social aspects like division of labour, cooperation, intervention of public and private agencies, processes and approaches followed by the members for the dissemination of technology, and extension approaches, etc.

INTERVIEWS

Interview was another most important tool for data collection. Both structured and non-structured interviews employed for the collection of data from farmers, researchers, and NGO workers. Some of the important questions were asked in order to understand the farmers' views on soil, water, land, forest, and their association with the agriculture and technological adoption, on farmers' rationality in selection of tools, seed varieties, production of particular crops, land selection, use of irrigation, etc. Various questions were put to understand the problems and possibilities in adopting modern tools and techniques, use of chemical fertiliser, HYV seeds, irrigation, marketing mechanisms role of public and private agencies, etc. Some of the significant issues were derived by asking questions such as how the changing economic policies have affected their production, and selection and adoption of different technology.

SCHEDULES

Household schedule was helpful to understand the socio-economic situation of the family in particular and the community in general. The population, educational background of the family and community as well were understood with the help of the schedules. The information regarding the income of the family, total land, types of land, and livestock possession were collected with the help of household schedule. Apart from household schedule, schedule for farmers and schedule for agricultural officials were also employed to collect data.

PARTICIPATORY TECHNIQUE

Participatory technique was useful not only in collection and understanding various aspects but also in cross checking the data collected through traditional approaches. Some of the important tools are as follows:

SEASONAL DIAGRAM

It was helpful to collect information with regard to seasonal cropping pattern, crops harvested, consumption, MFP collection, marketing mechanism, festivals and rituals, etc.

MATRIX TOOLS

Different matrixes such as crop matrix and institutional matrix were used for understanding and analyzing the reasons, assessing the problems about the crops grown. It also helped to understand in evaluating each variable. Similarly the institutional matrix was helpful to judge the value and importance of each variable of the respective institutions in sequence.

SECONDARY DATA COLLECTION

Secondary data collection was from the old and new documents available in various libraries, offices and institutions. Collection of secondary information included the loans received by the farmer from each bank, for each crop or economic practice, about the types of schemes allocated to each beneficiary. The types of projects launched, the allocation of budget in irrigation, soil conservation, HYV seeds, etc.

SIGNIFICANCE OF THE STUDY

The study is significant mainly from three angles. First, it is an attempt to understand the issues related to technological adoption in agriculture in tribal areas. Secondly, focus on the rationality, knowledge and innovation provides new inputs to the understanding of development of agriculture. And thirdly, analyses of the intervention of public and private agencies and new economic policies and their impact on agriculture in tribal areas are significant for the developmental policies and plans.

❑ ❑ ❑ ❑ ❑

CHAPTER - II

PROFILE OF THE STUDY VILLAGES

KANDHAMAL

Kandhamal, literally known as 'Kandha Terrain' is situated in the hilly regions of central Orissa. Famous anthropologist F.G. Baily during 1950s came to this district, the then Phulbani, as a researcher and focused the worldwide attention. "Etymologically, Kandhamal signifies the mountainous nature of its landscapes as well as dominating presence of Kandhas" (Jena, 1999, 154). This is one of the centrally located districts in Orissa. The Kandhamal is an extension of the Eastern Ghats at the point where those mountains terminate on the southern bank of the Mahanadi River. It is situated between 1,500 and 1,700 feet above the sea level. It is bounded by Ganjam district in the southeast, Rayagada and Gajapati in thc south, Boudh district in the north and northwest, and Nayagarh district in the northeast directions. This district is girdled by precipitous mountains, which separate the area sharply from the plains around. The hills are covered around the district and most of the hills are covered by dense forests and a few of which are bare due to the excessive use by woodcutters. The most common species are sal (*Shorea robusta*), mahua (*Bassia latifolia*) and mango (*Mangifera indica*). These trees are important in the economy of the people and the products collected are marketed by the people to support their economic system.

Population

Kandhamal, as the name signifies the 'Kandha Terrain', is inhabited by mixed group of communities. Historically, this place has attracted different groups of people from different regions because of its pleasant climate, natural beauty, clean environment and rich natural resources.

As per 2001 census, the total population of the district is 6,48,201 with decadal growth rate of 18.60%. The percentages of male and female are 49.80% and 50.20% respectively. The total number of tribal population is 336809 (51.96%), scheduled caste 109506 (16.89%) and others 201886 (31.15%). It is a district numerically dominated by the scheduled tribes. The scheduled tribes consist of Kandha, Gond, Kutia, Dongria, etc.

Table 2.1: Population distribution in Kandhamal district

Social Group	*Male*	*Female*	*Total Population*
ST	166283 (51.51%)	170526 (52.40%)	336809 (51.96%)
SC	54417 (16.86%)	55089 (16.93%)	109506 (16.89%)
Other	102099 (31.63%)	99787 (30.67%)	201886 (31.15%)
Total	322799 (49.80%)	325402 (50.20%)	648201 (100%)

Source: District Census, 2001.

Geographical Location

The district lies within the deccan plateau to the west of eastern ghats. Therefore, its climate is largely akin to the deccan region. It is subtropical, characterized by hot and dry summer/ cool and humid monsoon and cold and dry in winter. It lies between latitudes 19°34" and 20°54" north and longitudes 83°30" and 84°-48" east. It is located 300 m to 1100m above the sea level.

The temperature varies between 40°C and 12°C. The average annual rainfall is 1587m.m. Total forest area in the district is 4.39 lakh ha. Of the total cultivable land, 1.43 lakh ha are plain land, medium land covers 0.21 lakh ha, and low land covers 0.11 lakh ha. Net irrigated area is 0.18 lakh ha.

Soil Type: Most of the area of the district is red laterite group of soil (red sandy) which is light textured, porous and acidic in nature having ph 5.3 to 6.5. The soil is very porous with low moisture retention capacity and subject to heavy run-off and soil erosion during *kharif* season. G. Udayagiri block of Kandhamal district mainly constitutes red and yellow soil (Alfisol), and moderate soil moisture content.

Administration

Historical Background

The present Kandhamal district was an integral part of Boudh from time immemorial till 1855. As literature depicts, the earliest history of this area is gleaned from a number of copper-plate inscriptions issued by the kings of the early Bhanja dynasty that reigned over Boudh and Kandhamal in the 8th and 9th Century. Their kingdom was known as Khinjali Mandala. From the 10th Century to the advent of British in this region, Boudh, including Kandhamal, has been governed in succession by the following royal dynasties: the Somavansis, the Chindak Nagas/ Telugu Chodas, the Kalchuris and the Bhanjas. The history of Boudh-Kandhamal for 500 years prior to the coming of the British is however, still nebulous[1].

The Britishers launched a vigorous campaign in these hilly tracts with the objectives of annexing the areas to their empire and suppressing the abominable practice of human sacrifice, then prevalent among the Kandhas. The Britishers encountered stiff resistance from the tribals for a prolonged period of 20 years from 1835 to 1855. As the Boudh Raja utterly failed to curb the horrendous ritual of the tribals, the British truncated a large area, where the Kandhas were predominant, from Boudh on February 15, 1855 and named this newly annexed territory as Kandhamal.

After British conquest of Uttar Ghumasar (G.Udayagiri area) and Uttar Khemundi (Balliguda area) these territories were placed under the administration of the Collector of Ganjam district. These areas remained under the control and administration of the British until India attained independence.

Kandhamal remained a tahasil from 1855 to 1891 and it was administered by a Tahasildar under the direct control and supervision of the superintendent of the Tributary mahals of Cuttack. In 1891, it was upgraded to sub-division and tagged with Anugul district. When the new province of Orissa was formed in 1936, and Ganjam was merged with Orisaa, from the Madras presidency, Kandhamal became a sub-division of Ganjam. In the wake of the amalgamation of the princely states with Orissa in January 1948, Boudh and Kandhamal constituted the new district of Boudh-Kandhamal, with its headquarters at Phulbani. Balliguda sub-division was added to Boudh-Kandhamal district on 1.1.1949. With the separation of Boudh from Phulbani district as a new district, only Balliguda and Kandhamal sub-divisions remained with Phulbani district, which was later rechristened as Kandhamal in June, 1994.

Since tenth century the surrounding feudal states periodically tried to assert their sway over Kandhaland. The Kandhas resisted outside political domination till the seventeenth century. This was when the Oriyas began to settle in fortified villages within Kandhaland and cultivated on irrigated land around the settlements. To facilitate administration, the area was divided into muthas (10-15 villages) under the control of headman who were supposed to pay a tribute to the raja. In course of time, however, the mutha heads retained only the loosest of ties with the rajas of the plains. In 19th century when the area was ruled by the British administration, the mutha head agreed to pay the fixed amount as tribute to the British government through the rajas (Pathy, 1984).

However, the post British period was completely different. The muthadar system didn't continue for longer time. In 1960s the government removed the muthadar system. As far as the general administration is concerned, the entire district was governed by the state and rules and administrative procedures were followed as per the government.

Administratively the district constitutes with two sub-divisions, 12 blocks, 153 GPs and 2515 villages. Physiographically the entire district lies with high altitude zone with inter-spreading inaccessible terrain of hilly ranges and narrow valley tracts which determines the socio-economic conditions of people and development of the district. Overall, the district is ranked as a backward district in the state of Orissa.

The present Kandhamal district is made up with some segments of three erstwhile principalities of Boudh, Ghumsar and Khemundi ruled by the Bhanjas and the Gangas from ancient times. Their reign came to an end when the British came to this region in the nineteenth century.

G. UDAYAGIRI

G. Udayagiri constituted the northern fringe of Ghumusara kingdom of the Bhanjas. They occupied this state in the 9th century and continued to rule over it till 1835. Ganjam came under the Britishers in 1765. The Bhanjas could not put up with their interference and aggressive attitude from the very beginning and they raised the banner of revolt frequently against the British. The Kandhas and the Paikas forming the Ghumasar army waged relentless wars under the able leadership of Dohara Bissoyi from 1815 to 1835.

Deposing 'Dhananjay Bhanja' for his habitual revolt the British occupied Ghumusar on November 3, 1835. Dhananjaya Bhanja died at G.Udayagiri in December of the same year as a fugitive.

Health

Health is an important indication of social development. Despite its importance for the measurement of development index, the district does not show the expected progress even after decades of the implementation of health programs. The district has the lowest human development index (HDI) in the state. The district has the highest Infant Mortality Rate (IMR), which is 169. The human development index is 0.006, the lowest in state. Malaria, diarrhea, skin diseases are common diseases among the inhabitants in the district. The health facilities are not as much developed even five decades after the independence. Poverty, lack of awareness, and sanitary problems are the other health related problems in the district.

Literacy

As far the literacy of the district is concerned, the district has achieved some progress in the last few decades. The percentage of literacy as per 1991 census was 37.03% only. And, the tribal literacy percentage as per 1991 census was only 27.49%. However, as per 2001 census the percentage of literacy is 52.95%. The percentage of male literacy is 69.98% and the female literacy is 36.10% respectively. So there is substantial progress made in the literacy achievement of the district.

Table 2.2: Distribution of literacy percentage in the district

Census Year	*Total Male*	*Total Female*	*Total*	*Scheduled Tribe Literacy*		
				Male	*Female*	*Total*
1991	54.68	19.82	37.03	38.32	9.21	27.49
2001	69.98	36.10	52.95	NA	NA	NA

Source: District Statistical Hand Book, Kandhamal for the year, 2001.

Poverty

Though the district is known for its abundant natural resources, still the district lags behind of other district in its eradication of poverty. The district-wise poverty ratio of the State shows that Kandhamal has 75.42 per cent population under below poverty line. That means it is one of the poorest districts in Orissa and India as well (Das, 2005). The worst affected of the district are the scheduled tribe and scheduled caste population.

The lack of irrigation support in the hilly terrains, inaccessible villages, and issues of heath and education are some of the reasons for the backwardness of the district.

Forest Economy

Most of the people in the district are landless labourers and depend upon forest and other minor forest products. The economic importance of forest produce in the district is bamboo, kendu leaves, hill brooms, tamarind, sal seeds, siali leaves, etc. About 20-25% of family income comes from this enterprise. There are cooperative societies such as Tribal Development Cooperative Corporation(TDCC), Large Sized Multipurpose Cooperative Societies (LAMPCS), and Agency marketing cooperative societies (AMCS) having track records of marketing of minor forest produce. However, with the passage of time some agencies such as LAMPCS and AMCS have stopped involving in minor forest produce (MFP) procuring activities. With the enactment of Panchayat Extension to Scheduled Areas (PESA) Act in 2001-02, the Panchayats have been empowered with the selection of dealers from their own region. Now, there are private dealers in the district for procuring MFP.

Wage Labour

About 60% of the total workers in the district are agricultural labourers and other workers. Since job employment is limited in thc district many people go on migration to seek job opportunities elsewhere. Substantial family income of 40-50% comes from wage earnings. The statistical outlay of Orissa (2003) shows that there are 1,02,380 cultivators, 1,10,190 agricultural labourers, 21,588 workers in household industry, and 71,889 other workers in the district.

Agricultural Scenario

The district lies in the eastern ghat agro-climatic zone. The district positioned number-1 on the spices production in the State.[2] Apart from the spices, there are some regions like Phiringia, G Udayagiri, part of Phulbani block etc. in the district have well progressed in the vegetable production. The major vegetables such as cabbage, cauliflower, beans, brinjal, and potato have attracted the businessmen from the other districts. Cereals and pulses production are also important part of the agricultural yield in the district. Mostly, the agricultural technologies are traditional types. Use of bullock, wooden plough, and other traditional implements is the significant feature of the traditional agricultural technology. Apart from that, the district has got the wide attention of the private and public agencies due to very less application of the chemical fertilizer and pesticides and insecticides. During last decades, there is increase in adoption of modern technology which is discussed in the next chapter.

Operational Landholding

The total number of operational landholders in the district is 83,922 with operational holding of 1,06,771 ha and average landholding of 1.27 ha. Small and marginal farmers constitute 81.8% of land holders having command over 52.6% of area while the big farmers constituting 18.2% of total landholders are in possession of 47.4% total area.[3]

Table 2.3: Distribution of operational landholding in the district

Sl. No.	*Particulars*	*Unit*	*No/ha*
1.	Operational holding Marginal (<1ha) Small (1-2ha) Semi-Medium(2-4ha) Medium(4-10ha) Large(>10ha)	No	83922 44707 23969 12273 2767 206
2.	Area owned Marginal farmers Small farmers Semi-medium farmers Medium farmers Large farmers	Ha	106771 22677 33449 32553 14434 3658
3.	No. of cultivators	No	102380
4.	Agricultural labourers	No	110190

Irrigation Potential

The irrigation potential so far created in the district can irrigate 12% of cultivated land during Khariff (16093ha) and 7% area during Rabi (5965ha). The actual area irrigated is reported to be less than the potential area. Of the gross cropped area of 1,66,051 ha during 2001-2002, 22,257 ha was irrigated which means 13.40% of cropped area enjoyed irrigation from all sources while 86.50% of cropped area remained rain dependent.[4]

The Kandha

The word 'Kandha' is spelt variously which are synonymous such as Kond, Khond, and Kandha. The language they speak is Kui, which has no script.

The 'Kandhas' are identified from their names. Some writers have attempted to trace out the Telegu derivation from the word Konda means hill. Those living on the hill tops are named as Kandha. It is a fact that the Kandha like to live in hill tops and their subject people the Panas like to live beneath their settlement. The common surnames of Kandhas are Pradhan, Mallick, Konhar, and Majhi. And religious functionaries have surnames like Dehuri, Jharkar, Jani etc.

ETHNOGRAPHIC RECORDS

Different views have been given by different authorities about identity of Kandhas:

- Dalton (1878) described the Kandhas as tall as average Hindus and much darker in complexion.
- McPherson (1847) described the Kandhas as faithful to friends, devoted to their chiefs, resolute, brave, hospitable, Laborious.

The Kandhas have their loyalty to their erstwhile feudatory chiefs in Orissa and elsewhere. They are treated as valiant warriors and discharged their services very faithfully to their rulers. They offered their valuable services at the time of freedom movement. To name a few among them are Chakara Bisoyi and Dohra Bisoyi.

Types of Kandhas

According to the area of habitation Kandhas are classified into three classes namely:

- Kutia;
- Malua or Dongoria; and
- Desia .

The Kutia Kandhas are found mainly in Kotgarh, Tumudibandh and Belghar area of Balliguda Sub-division. The Dongoria or Malua live in high lands of hilly area in the district. The Desia or Oriya Kandhas live in plain areas with the non-tribal.

Dress

Many studies as discussed by the anthropologists (Bailey 1957; Barbara Boal 1995) the Kandhas have followed the similar dress pattern like the Oriyas and plain inhabitants. However, there is a little difference in the dress pattern. The men wear a long and narrow cloth covering at least the knee portion. Young women use the hand made sarees, blouse and others and follow the plain Oriya dress pattern. Hair is tied with colorful ribbons. But the elderly and middle aged women wear sarees but very rarely wear the blouse. Comb and Hair pins made up of metal are used by them.

They bore the entire rim of the ear with silver rings. They tattoo their faces before marriage. However, the tattoo culture is gradually disappearing from the Kandhas who are basically coming across urban influence.

Food Habits

Food habits among the Kandhas are somewhat different from the plain area people or non tribals. They prefer gruel of rice mixed up with vegetables such as Lablab beans, French beans, *Papaya*, etc. which are commonly available in the region. They are fond of meat and fish. However, they eat them rarely.

They smoke and chew tobacco leaves. Both men and women consume excessively *Salapa* and Mohua liquor on all special occasions.

Houses

The Kandha houses are made up of wooden walls plastered with mud. The floor is of mud but well shining like black coloured marbles. The houses are thatched with date palm leaves, straws or of forested grass. Doors are made up of bamboo splits designed artistically.

Rituals in the Society

Child Birth

The pollution in connection with child birth ends on the fifth day. On that day father of the child sacrifices a fowl and offers cooked meat, rice and liquor to the ancestors. This is performed with the belief that no ill may befall the child. Some households perform this ceremony on the 7th day. After one month, hair on the head of child shaved off and a feast is given to the neighbours.

Death Rituals and Kraha Nata

Krahanata is an important dance performed during the death rituals. The Kandhas perform rituals to get purification after three or seven days of a sibling's death. This purification process is called as *mada*. On the day of *mada*, the relatives of the deceased gather at a place and perform *Sokasabha*. After the *sokasabha* they lick a special oil which is usually prepared for purification. On the day of *Sokasabha* the grandsons and grand daughters of the deceased member perform Krahanata in order to create a pleasant environment. The grandchildren wear with bullock or cow horn. Both male and female take part in the dance. The male individuals hold *tangi* (axe) and some beat drums. Matching with the musical rhythm both male and female dance in a different position.

Festivals

Kedu Festival

Kedu festival, which was once considered as a prime festival of the Kandha is now rarely observed. This is celebrated when the Kandhas feel that the mother earth (*Dharani Penu*) seeks the blood. When the earth becomes dry, Kandhas feel the trembling of the earth's surface, springs flowing on the surfaces get dried, the Kandhas call a meeting to immediately organize the festival. This festival is observed with the presence of all village people and invitees from the other villages. A buffalo is arranged or offered by a devotee to be sacrificed. Before few days of sacrifice, the animal is fed enough and all the people serve it through out the days. On the day of sacrifice the animal is tied to the pole, specially meant for sacrifice. Some elderly women chant the mantras and read the hymns. The *Dehuri* does the ritual performance and then the animal is sacrificed. After the sacrifice the people collect the blood and sprinkle over the soil in agricultural field and the mountains.

Though this belief is gradually disappearing, but it is celebrated once in five or six years with the belief that both social and economic security will not be hampered and they can avoid the external threats.

Religion

Kandhas are mainly animistic. They believe in soul and spirit. But gradually with the contact of plain Oriyas in the region, majorities of the Kandhas have embraced Hinduism. Similarly, the entry of Christian missionaries in the region has influenced the Kandhas to embrace Christianity. In G. Udayagiri, Phiringia, Raikia, Tumudibandh, and Daringibadi regions of the districts have quite large number of Christian population.

The Kandha are well known for the *Mariah sacrifice,* which was practiced a century back. They have strong faith on natural and supernatural entities. They believe that supernatural entities are associated with the success or failure of any deeds. They worship the nature in the form of mountain, soil, forest, etc.[5] The Kandha appease them by sacrificing animals and offering other foods and flowers. As a result, the social structure of the Kandha society is an intrinsic part of the configuration of man, nature, and supernatural entities.

Body Culture

Tattooing is one of the interesting cultural traits among the Kandha. The Kandha women exhibit the symbol of clan object, totemic object or the symbol of deities on their body. This culture is very old and is inherent since time immemorial. But this is gradually disappearing with the rise of modernization. Education, particularly higher education among the community members is also responsible for the same.

The Kandhas' Economy

The Kandhas are economically backward as compared to their counterparts in the district. Most of them are landless labourers or marginal landholders. Still, a bulk of the Kandha population depends upon forest and other minor forest products. The forest produce in the district are mainly bamboo, kendo leaves, hill brooms, *siali* leaves, *sal* seeds, tamarind, *mahua*, etc.

The Kandha adopt both shifting and settled agricultural practices. However, they are primarily agriculturists. Besides agriculture, their livelihood is based on collection, processing and marketing of MFP and a few of them do government jobs. Horticulture, spices production, sericulture and vegetable cultivation are some of the new economic initiatives in the district. Economic development has been given primary importance with the extension of ITDAs, LAMPS, DRDA and other cooperative and credit agencies.

The Panas

The Panas are mainly distributed in Balasore, Cuttack, Puri, Dhenkanal and Ganjam districts of Orissa and in the adjacent states of West Bengal and Bihar. The Pana have some endogamous occupational subgroups like Betropano, Buno Pana. Pana is a synonymous of Pano, Panwa and Pana. The Pana are divided into a number of lineages which regulate marriage alliances and indicate one's ancestry. They have only gotra, Nagasya. Traditional and primary occupation of the Pana is to serve as serfs to the landholding communities. However, a few of them are now engaged in service, business, daily wage labour and cultivation, and as musicians. Women mostly work as labourers (Singh, 1998, 2731-32). Nearly 100 years ago O'Malley[6] observed: "In the Kandhamals, the Panas were the serfs of the Kandhas. They worked on their farms and wove cloth for them, in return for which they obtained a small area of land, grain for food and all their marriage expenses; they used also to procure victims for the meriah sacrifices. Their serfdom was so well recognized that if a Pana left his master and worked for another, it caused serious dissensions among the Kandha community. To this day there is a settlement of Panas- a kind of Ghetto-attached to every large Kandha village, where they weave the cloths, the Kandhas require and work as farm labourers."

The Panas worship many gods and goddesses. Some worship Goddess Durga, Mahakali and other local deities. Baraladevi is the famous deity in the region. In district gazetteers it is mentioned that they also worship Dharni Devta. The people of this caste also participate in the Durgapuja, Diwali and Ratha Jatra festivals. Due to influence of Christian Missionaries a considerable number of Panas have embraced Christianity in Kotagarh, Tumudibandh, Baliguda, Nuagaon, Phirngia, Daringibadi and G. Udayagiri regions of the district.

The Pana is a scheduled caste community dominant as a single caste majority in the district. Since centuries, they became an inalienable part of the district. Most of the Panas in the region are landless laborers and wage earners. There are many who work as intermediaries between Kandhas and the *Sahukars* or *Mahajans* in the region. Since time immemorial, the Panas have influenced socio-economic sphere of the society.

The Kandha Pana Relationship

As far as the caste-tribe relationship is concerned, the Panas will be the first who are very closely connected to the Kandha communities. Panas are very friendly in nature. Their putative kinships like mahaprasad, maker, maitra, sai, and sadu exists with the Kandhas. Most of the Pana settlements exist in the Kandha settlement regions. They have secured their place along with the tribal communities for number of regions, which are as follows.

Economic Relationship

The economic relationship is very important when the Kandha-Pana relationship is concerned. The Kandhas are their clients both directly and indirectly. They often maintain their livelihood by depending mostly on the Kandhas. The production relationship is an important one. The Kandhas are basically primary agricultural producers. Where as the Panas are most of the times the sellers of the primary agricultural and minor forest produce collected and harvested by the Kandhas. The Panas know the trick of business. They collect the produce directly for the tribal people and sell it in the local-regional market. In other sense, they collect the produce from the tribals and sell them to the *sahukars* or *mahajans*. Similarly, the mahajans, who are belonged to non-scheduled caste communities, engage them as intermediaries to collect the agricultural produce seasonally from the Kandhas and dispose them at the *Godam* (Store house) owned by the mahajans.

Social relationship: During the time of community meeting of the tribals, the Panas work as the messengers. They beat the drum, make the people alert and spread the massage regarding the meetings. Similarly, during the time of death rituals and purification, death ceremony time, the relatives of the Kandhas are called with the help of Panas. This cohesive relationship exists in the form of *jajaman* and *kamin* relationship since time immemorial.

Again as far as the religious relationship is concerned, the Kandhas and Panas have both solidarity and discordance. The

Kandhas who are converted into Christians consider the Pana, the converted Christians, as their brethren. There is instance of inter-community marriage existing between Pana boy and Kandha girl. However, the marriage between Kandha boy and Pana girl is rare. The people who have embraced Christianity celebrate Christmas with great fervour. In general, the Christmas celebration is enjoyed by all communities across religion and caste. They celebrate Christmas together with great fun and care. They come together at the time of necessity.

In contradiction to this, the relationship between Pana and Kandha has bitter relationship in some cases. The Kandha, who have embraced Hinduism, consider the Panas as untouchables. By and large, the Panas are not allowed to enter into the formers' house. Some Kandhas dislike them because the Panas are viewed as exploiters. The Panas deceive them in business and other economic activities.

Despite all this, the Kandhas and Panas have great solidarity. They are good example of social harmony. In many tribal inhabited regions, though the population of the Panas is less, their influence to each other's culture and economy is important.

The Villages

Location of the Villages

The villages selected for the study are Sudhipara and Laburi. They are located in the G.Udayagiri tahasil, which is also a Block of Kandhamal district. The villages are centrally located between G. Udayagiri and Raikia towns. Raikia, another block of the district, is also a commercial centre. As a result, the people access this commercial centre for their purpose. The villages are surrounded by towns, villages and forest from the four directions. Towards south, there is Raikia block; north there is G. Udayagiri town. North-East is followed by the famous Kalinga *ghat* and then comes the Ganjam district border. In that way, the place lies at the centre surrounded by fringes of hills and dense forest as well as small towns. The place is no way completely isolated. Different groups of people are their neighbours, hill ranges are their boundaries, and forests are their resources. There are small springs flowing across the villages. Green vegetation and deforested hill patches is again the important feature of the villages.

Importance of the Villages

These two villages have a special recognition in the entire district from various points of view. First of all, this zone lies on the highest point above the sea level, which is almost at the same height where the Daringibadi block of the same district is located. Daringibadi is known as the second Kashmir in Orissa because of similar climate and sometimes even snow falling during the winter. Similarly, the region has somewhat similar climatic situation. Secondly, the farmers of the region are well known for the beans cultivation. Third, the adoption to new technology by the Kandha farmers is prompt as opined by the common people in the district.

Settlement Pattern and Household Composition

Housing Pattern: Almost all the households in the region are having asbestos roofing and rarely thatched roofing. More than 50% of households are made up of brick and cement plastered walls. Less than 50% households are having mud brick walls or sometimes with only mud. Floorings are of the same nature, which are either made by cement or by mud. A few traditional houses are seen among the elderly people who do not have descendants.

The change in housing pattern has become possible with change in occupation, governmental support, political affiliation and other factors.

The villages consist of 11 hamlets of which Sudhipara village has seven hamlets and Laburi village has four hamlets. The two villages are connected by both pucca and kacha roads. Where as, the hamlets of each village is diagonally as well as horizontally connected through kacha roads. The settlement pattern is very scattered. These two villages are mainly inhabited by the Kandhas and the Panas.

The villages consist of 123 households. Out of which, the Kandhas consists of 91 households, Panas 30 households and others have only two households. In Sudhipara village, there are only six Pana households where as Laburi village is having 24 Pana households.

Table 2.4: Household details

Villages	*Hamlets*	*Kandhas*	*Panas*	*Others*	*Total*
Sudhipara	Adunaju	5	0	0	5
	Bijigam	6	0	0	6
	Kala Sahi	4	3	2	9
	Kupanaju	4	0	0	4
	Madinaju	7	0	0	7
	Samarada sahi	14	1	0	15
	School sahi	22	2	0	24
Laburi	Kepad paderi	18	0	0	18
	Kilakia	5	6	0	11
	Laburi	6	18	0	24
	Madinaju	0	11	0	11
Total		**91**	**30**	**2**	**123**

Population

Mainly the villages are numerically and economically dominated by the Kandha. The total population of the village is 563; of which 432 are Kandhas, and 129 are Panas. In addition, there are only 2 members of general caste population. The male and female population ratio in both the Kandha and Pana communities is almost same. Among the Kandha, the population of male and female consists of 221 and 211 respectively. Among the Panas, the male and female populations are 69 and 60 respectively. The age-wise population for both the Kandha and Pana are given in Table 2.5.

Table 2.5: Population distribution

(I) Age and social group-wise distribution

Age Group	*Kandha*		*Pana*		*Others*		*Total*		*Grand Total*
	Male	*Female*	*Male*	*Female*	*Male*	*Female*	*Male*	*Female*	
0-10	45	51	21	12	0	0	66	63	129
10-20	52	38	10	18	0	0	62	56	118
20-30	37	35	10	7	0	0	47	42	89
30-40	37	36	12	9	0	0	49	45	94
40-50	25	16	6	8	1	1	32	25	57
50-60	13	23	6	4	0	0	19	27	46
60-70	10	11	4	2	0	0	14	13	27
70-	2	1	0	0	0	0	2	1	3
Total	**221**	**211**	**69**	**60**	**1**	**1**	**291**	**272**	**563**

Household size: The household size varies from household to household. The average household size is 5.

Family structure: Family structure is mainly joint family type. Among both Panas and Kandhas, the son along with his wife and children stays with his parents. Nuclear family is also an important family type in the villages. The property of the father is shared by his son but not by the daughter. The daughter after marriage lives separately and does not demand her father's property.

Occupational Details

The occupational details of the villages show that agriculture is the primary occupation of 83 households which is 67.5% of the total households. The other 40 households have different primary occupations. Wage earning is the second largest primary occupation in the village. As far as secondary occupation is concerned, wage earning ranks number one in the villages with

53 households followed by business and other activities. The table given below shows the details about the occupational distribution among the households. The other significant observation that is noticed from the villages is the migration of some people (3 No.) to Kerala for employment opportunities in industry or related field to support their livelihood. During recent years there is drastic decline in the MFP collection. The decline of forest resources due to over population is also a crucial factor for the loss of occupation resulting in migration.

Table 2.7: Occupational distribution (Primary occupation)

Sl. No	*Occupation*	*Households (No)*	*Percentage*
1.	Agriculture	83	67.5
2.	Wage earning	15	12.2
3.	Business	7	5.7
4.	MFP collection and other	6	4.9
5.	Service holders	4	3.3
6.	Migrated laborers	3	2.4
7.	Others	5	4.0
	Total	**123**	**100**

Table 2.8: Occupational distribution (Secondary occupation)

	Occupation	*Household (No)*	*Percentage*
1.	Wage earning	53	43.1
2.	Business	16	13.1
3.	Agriculture	13	10.6
4.	MFP collection	11	8.9
5.	Carpentry	1	0.8
6.	Other	29	23.6
	Total	**123**	**100**

Tables 2.7 and 2.8 show that the agricultural occupation is predominance in the region with 96 household having this as primary and secondary occupation. This shows that more than 80% households depend on this.

INCOME DISTRIBUTION

The income distribution among the households shows that there is large gap between rich and poor households. The income distribution varies from Rs. 1 lakh per annum to less than 10,000 per annum. Some households having land and employment in government sector comes in the income category of Rs.1 lakh and above. The total households under this category are 04. Similarly, the landless households who are particularly having no option other than MFP collection come under less that Rs. 10,000 per annum category. This category includes mainly the households having only elderly people and do not have children or having children are living separately. Fifty-one households (around 41%) have average income between Rs. 40, 000 and Rs.60, 000, per annum which varies again on the basis of yielding from agriculture and availability of wage earning.

Table 2.9: Annual income distribution of the households

Annual income	*Kandha households (No)*	*Pana households (No)*	*Others (No)*	*Total Households (No)*
< 10000	10	05	00	15
10,000-20,000	09	06	02	17
20,000-40,000	11	06	00	17
40,000-60,000	39	12	00	51
60,000-80,000	12	01	00	13
80,000-100000	06	00	00	06
>1,00000	04	00	00	04
Total households	91	30	00	123

Similarly, 19 households have income of around Rs. 60, 000 to 80,000 per annum. Again, the above table shows that there is no family from Pana households having more that Rs. 80,000 annum income category. However, there are five households out of total 30 Pana households having less than Rs. 10, 000 incomes per annum.

The above discussion on occupational distribution and income distribution plays significant role in the adoption of technologies, choice of crops, and production practices in agriculture. The detailed discussion about it is discussed in the following chapters.

Education and Literacy

There is a Primary School in Sudhipara village for the elementary education of the children. This school was set up in 1964. Since that time, the people have accessed the opportunities of education. Similarly, there is one primary school located in Laburi village. There is one High School called Mahatma Gandhi High School located on the mid way between Sudhipara and Laburi villages. As far as school education is concerned there are better facilities for the people in the region. For higher education, there are colleges located in Raikia and G. Udayagiri towns which are 15 kms each from the studied region.

Literacy

with regard to literacy, there is a large gap between male and female. Of the total 310 Kandha literates in the villages, the literacy among male is 188 where as the female literates are 122. Among Panas, out of total 107, the gender wise literacy is 64 male and 43 female. In general, the literacy rate in the villages is quite satisfactory, which is a little higher than 74%. As far as the Kandhas are concerned, the literacy rate is very satisfactory. Surprisingly, these villages have achieved substantial progress in literacy in comparison to the literacy percentage of the district which is below 52%.

Table 2.10: Educational Distribution

Educational Qualification	*Kandhas*		*Panas*		*Other Male*	*Other Female*	*Total*
	Male	*Female*	*Male*	*Female*			
Primary	75	42	23	16		1	157
Minor	60	34	18	19		0	131
Secondary	34	18	12	3	0	0	67
Higher secondary	7	7	3	2	0	0	19
Graduation	9	3	1	0	0	0	13
Post-Graduation	3	0	0	0	0	0	3
Total population	188	104	57	40	0	1	390
Illiterate	22	88	5	17	1	0	133
Total	**210**	**192**	**62**	**57**	**1**	**1**	**523**

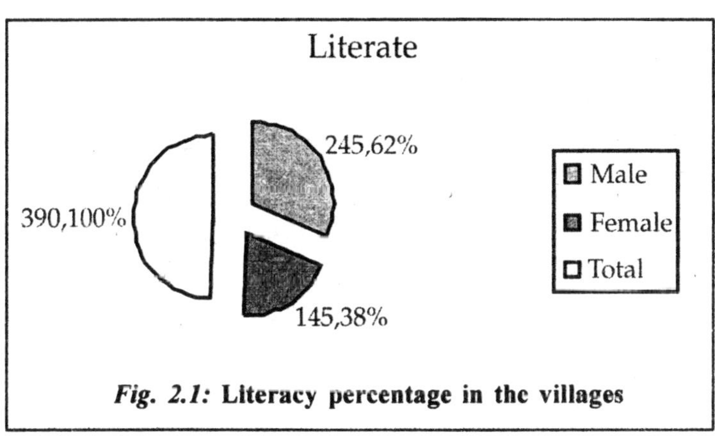

Fig. 2.1: **Literacy percentage in the villages**

But the major problem is that the skewed literacy distribution between male and female. When the male literacy is 245, the female literacy is 145 only in the villages. This consists of 62% male literacy and 38% female literacy.

Health

Malaria, diarrhoea and dysentery are the common diseases in the region. Due to lack of adequate health facilities, the region has not shown any progress to achieve health improvement. Malaria eradication program has not been successful due to irregularity of the ANM coming to the villages. Similarly, there is no primary health centre in the nearby region. A private health centre is opened nearby Sudhipara village, which is the only source for immediate heath care. During the time of mango or jack fruits ripening, the people suffer from diarrhea and dysentery for over consumption of the unclean and un-hygienic fruits collected form the road side. The main victims of this are the children. According to the people, no awareness creation has been made from government for heath care so far. Anganwadi worker (AWW) provides counseling and supplementary diet to pregnant and lactating mothers. She also keeps track of expected mothers by taking their weight weekly and advises them on good practices during pregnancy. She provides supplementary diet and pre school education to children.

There are a couple of quacks practicing in the villages and one registered medical practitioner is present. People usually prefer to visit to quacks or RMP in case of minor medical problems. Only in case of a serious matter they go to sub-divisional or district private doctors.

Infrastructure and Resources in the Villages

Road Facilities

There is a pucca road linking to both Sudhipara and Laburi villages which is straight connected to the main road (Phulbani-Daringibadi). This road was constructed as a part of the *Pradhan Mantri Gramya Sadak Yojna* (PMGSY) during 2001. Apart from the pucca road there are some treckings created by the villagers for time saving journey. Lingagada, which is 5 kilometers from the Sudhipara villages and seven kilometers from Laburi village, is the main local market for the local people. Therefore, for easy communication there was another bypass road being constructed from the Lingagada Panchayat to the Sudhipara village as a part of Food for Work Programme (FWP). The FWP program was replaced by National Rural Employment Generation Programme (NREGP) in 2006.

Electrification in the Villages

In 1990s the first electricity connection was given to the villages. There are 20 households in the villages having electricity connection. There is one rice meal running in Sudhipara village with the help of electric power connection. Use of dish-antennae is also observed in the villages. There are two households in Laburi having dish antenna for domestic purpose. As there is dish antenna and television in the village, people from nearby villages go and watch television there.

In Sudhipara village there is one television set owned by a Kandha family. Many people from the same villages often go there and watch serials and movies on television. Similarly, people eagerly see agricultural related programs on television. Supply of electricity has certain impact on the culture and agriculture in the villages.

Market Facilities

For the people of Sudhipara and Laburi, the main market centers are Raikia and G. Udayagiri. Raikia has attracted the people from different places from different regions due to the famous turmeric and ginger business hub. From nearby villages, people come to the centre during the *haat* days especially to sell and buy their agricultural produce. Similarly, G. Udayagiri town is an important place after Raikia as the distance is equal from the villages. Similarly this is the block and tahsil headquarter for the villages. For buying machineries, agricultural tools and implements the people depend on both these market places. Similarly, for buying traditional agricultural tools people come to blacksmiths who have opened the work places and selling centers at the corner of each town.

There are buses and lorries plying between these two towns regularly. So transport facilities are better off in the region. Many people go to these towns even by cycles also.

Other Institutional Details

There are some cooperative agencies and banks located in the region. They are: LAMPCS at Lingagada, SBI at Katingia, and fertilizer cooperative society at Raikia town for the cooperative and financial support and other input supply for the farmers in the region.

Forest and Forest Land

Forest is the major resource for all living creature. It is an inalienable part of every tribal community. Most of the tribal communities in India are surrounded by forest and are settled in the forested regions. The villages are not exception to it. The villages are surrounded by the hills. *Sal* trees are common in the region. However, some hills have become barren after frequent cutting of trees and some practice of podu cultivation. After the ban of forest cultivation, still there is no improvement in the forest cover. Now, the promotion of participatory forest management through self-help group formation has more or less able to restrict the exploitative use of forest by the insiders and outsiders.

Forest is an additional economic source for the communities inhabited in the region. Mostly, the landless laborers, marginal landholders depend on the forest for their livelihood. Collection of fuel wood is an important activity of each and every household in the region. Sal seeds and Sal leaves collection has been an additional occupation for more than one-third population of the district. For use of wooden agricultural implements people also depend on the forest.

REFERENCES

1. *Orissa District Gazetteers:* Boudh-Khondmals has an Elaborate Discussion About the History of Kandhamal, the then Phulbani District.

2. *Agricultural Census*, Government of Orissa, 2005.
3. Xavier Institute of Management, Study Report, Bhubaneswar, 2006.
4. *Ibid*.
5. Felix's (1995) book on The Sacrifice of Human being and Baol's (1997) Book on Human Sacrifice and Religious Chang are some of the Classic Ethnographic Collections on the Kandha.
6. L.S.S.O. 'Malley, Bengal District Gazetteers, Anugul, 1908, pp. 42 cited in *Orissa District Gazetteers*, Boudh-Kandhamal, 1983.

❑ ❑ ❑ ❑ ❑

CHAPTER - III

TRADITIONAL AGRICULTURE

Traditional agriculture, as an important background discussion encompasses Kandhas traditional agricultural practices in particular and the agricultural practices in the region in general. The discussion of the Kandhas' traditional agricultural practices is essential as the Kandhas are the original inhabitants in the villages as well as in the entire Kandhamal region. Traditionally the Kandhas are shifting cultivators. In later part, of course, the entry of Pana and other non-tribals influenced this traditional agricultural practice. But, the entry of Pana has not much affected the traditional practices. The chapter is aimed at understanding the traditional knowledge, wisdom, technology adoption, and crops production of the Kandhas.

The Kandhas are considered to be well known traditional agriculturists in Indian tribal history (Padel, 1995; Boal, 1997). They are recognized for turmeric cultivation. Both shifting and settled cultivation are practiced by them. Traditional cereals, pulses, millets and cash crops of the region attract a special attention of the outsiders. Moreover, agriculture among the Kandhas, historically, has significant association with their cultural milieu. For quite a long period, their beliefs in natural and supernatural objects, cosmology, perception and knowledge system dominated the process of agricultural production. They are nature lovers. The history of the tribe depicts that the Kandhas believe their origin to be from *dangar* (mountain) along with other plants and animals. They respect soil, mountain, forest and other natural objects as their deity or the mother. They call mountain as Saru Penu (mountain mother) because all the plants and animals originated from the soil. According to them, it is the mother earth that decides the existence and development of crops on its surface. The mother earth controls both rain fall and climatic fluctuation.

The Kandhas believe man alone cannot predict the climate. There are some animals and plants sent by the mother for them to predict the climate before the onset of a season. The Kandhas have identified several birds that alert the farmers to be ready for cultivation in the next season. According to them, if a bird sounds like *matatapu* it is the time to store the seeds for the next season. In the earlier days when the Kandhas, most of their time, remained intoxicated in consumption of liquor prepared out of date palm, sago palm, or *mahua* flower the bird only enabled them to know about the onset of the seasons. It was being believed that if the same bird sounds *"birinpiyadi"* it is the onset of monsoon. Traditionally, *rajabani* (a bird) enabled them to know about the times and about the auspicious days. However, in due course of time, this has become a story of the past. People are now more aware of the modern technologies with the mode of electronic media such as radio, television and other form of communication systems.

ORIGIN OF AGRICULTURE

The Kandhas believed that agriculture was introduced by "Saru Penu", the hill deity. Mother earth taught them how to grow crops. However, many writings depict that the Kandhas mainly knew about settled agriculture from the non-Kandhas (Baily, 1957; Thurston, 1902; Boal, 1997; and Pathy, 1984). But still many myths exist in the Kandha society. Kandhas belief that '*Dand Penu/Saru Penu* (hill mother)' had enabled them to know about ploughing, sowing, weeding, harvesting, etc. They know about the agricultural practices since time immemorial. It is believed that 'Dand Penu' had demonstrated the Kandhas about ploughing through an agent. The deity came in dream of the agent and demonstrated about ploughing. This was the initiation of ploughing among the Kandhas. Thereafter, this became a general practice. The Kandhas believe that the *Saru Penu* has her own animal called Sambar (a deer like animal) for ploughing purpose. This agricultural practise was inherited from mother through some spiritual agents and finally reached to the community.

HILL CULTIVATION (*PODU CHAS*)

The first of any agricultural practices among the tribals is the hill cultivation. According to them, the hill cultivation or *Bagad Chas* did not arise spontaneously. They believe that it was the handiwork of *Saru Penu*. Sometimes, the Kandha call it as *Podu Chas*, as the cultivation was done by slash and burn method. The crops that were produced from the podu cultivation included large varieties of millets, some cereals and pulses.

PROCESS OF CULTIVATION: LOCAL PRACTICES

Slash and Burn Method (Podu)

Before the onset of rainy season, the Kandhas cut the bushes and trees. They collect and accumulate them at a place to let them dry in the sunny days during February and March. These dried plants are set on fire after performing a ritual, called *Bagad puja*. All the villagers along with the ritual leaders assemble at the place of *bagad*. The ritual process is observed with the sacrifice of fowl or by offering eggs. Rice and turmeric are offered along with the fowl. The fowl is sacrificed on the spot and a few drops of blood are dropped on the wood gathered on the field. After the ritual, the ritual leader set on fire to the dried woods. After which, all the people assembled there take some pieces of charcoal or burnt end of the wood to add in the podu soil of their respective lands. This ritual is observed with the faith that the plants that grow on the podu land would never be destroyed by the natural and man-made events. Forest mother (*Gassa penu*) is worshipped to bring happiness for the seasons ahead with all her graces.

The Kandhas prepare the podu land in a particular direction and with perfect angle. The ploughing of land is made in a criss-cross manner in order to make the soil more porous and lighter. The ploughing of the lands is done both horizontally and longitudinally and it is based on the crops selected. The ploughing that is made in a longitudinal shape is called '*ekasiuhasu*' and the ploughing with horizontal shape is '*rasilakuhasu*'.

There is no ridge prepared on podu land. Water is allowed to flow down smoothly on the soil. The seeds are transplanted with caution so that it won't be washed away by the water flow. Soil erosion was less in this type as the field is surrounded by the forest. Each patch of land is separated and divided with the presence of the forest. Therefore, the land is otherwise called as forest land *(gassa ketanga)* or bagad.

SOIL TYPE

The podu land contains *kala bira and balu bira* (black and sandy soil). The soil type is porous, and the moisture content is normal through out the year. According to some Kandhas, even thirty years back the podu land was so fertile that a single plant of *kanga* (a pea variety plant) could even produce 5 kg of seeds. There was no harm to the soil either by irregular rainfall or sunlight. As the soil was porous and light, the water holding capacity of the soil was better. The soil could receive air. As a result the crops grew well on the podu land.

In order to maintain fertility of the soil, the Kandhas leave the land fallow for a period of 4 to 5 years. During the time of cultivation, the crop rotation mechanism is followed by the Kandhas. Mixed cropping was also another mechanism to protect the soil- fertility on the podu land.

CROPS CULTIVATED

Traditionally, the Kandhas used to cultivate varieties of cereals, millets, pulses and cash crops on the podu land. Cereals namely, paddy, *jower* and maize, millets namely, *kuiri, ragi and klingiraka* and spices like turmeric and ginger were being grown on the land. Podu land is suitable for growing up of above crops.

These millets were the main staple crops till 1970s. Millets and pulse varieties used to be cultivated on a rotation basis in the same land. If this year they cultivated kanga (Pegion pea) on the portion of podu land, the same land is used for millets in the next season. This rotation mechanism in cropping pattern formed main feature in traditional agriculture.

The traditional cropping pattern among the Kandha is mainly mixed cropping. No millet is cultivated in isolation. However, the seed of different millet varieties are sown with a limited interval of time in the following order.

Kuiri ———→ Ragi ———→ Arca ———→ Kangu

(In a limited interval)

The selection of these crops is done in such a way that the plants could be identified and differentiated from each other after they grown up.

The land is usually ploughed with the help of bullock. In March and April, people sow seeds. They wait nearly two months until the heights of the plants reach at least seven to eight inches. Weeding is done in *Asadh* (June-July) and *Sravan* (July-Aug). Mainly, women do the weeding activity. The people watch the crop field to protect them from the threats of the animal and this continues till *Aswin* (October-November) until the crops start maturing. The crops need six to seven months for a total cycle of production. First, a woman broadcasts *kuiri* seed. The broadcasting is done with the skill so that the seeds spread across equally all over the soil. Sometimes, it is replaced by *ragi*. *Arca* is broadcasted just after the *kuiri* but in a definite row or column to differentiate it from each other. However, all the millets are grown on the same field. The purpose to broadcast a variety of seeds in the same land is to maintain the fertility of the soil and for producing multiple millets from a specific cultivation procedure at a single time. They say that it was also done in order to protect the land from wild animals and birds. The limited cultivable land located at a particular place was easy to watch by a few people. Therefore, broadcasting the seeds in different lands at different location was not suitable from the security reasons as well as because of shortage of human forces.

Kanga or Kangul (Pigeon pea, a pulse variety)

Kanga seed is broadcasted on the soil in March and April. Seeds start germinating in May and June. The first weeding takes place in June. In a period of one and half months the plants grow to a particular size, which needs a second weeding. The plants start bearing flower in September and October. In November, generally the tender pods come out. The pods get matured in a period of three to four months time. Harvesting of Kanga is done after the observation of a ritual called, *kanganenju*. In this ritual, both men and women take part. Celebration of ritual is done with great fervour. The first fruit after harvesting is offered to the goddess earth. Like other rituals in the region, the ritual leader, *Dehuri*, offers flower, fruits, newly harvested kanga seeds and sacrifice fowl to the mother earth. After that, men celebrate by consuming liquor. This ritual symbolises the respect and obligation to the mother earth. Good production brings happiness among the Kandhas.

Turmeric Cultivation

Turmeric is the most significant cash crop in the region. This is locally called as *sidiganga*. Kandhas have been cultivating it since time immemorial. There is no evidence about the exact beginning of the turmeric cultivation in the region. It is believed that the origin of turmeric date back to Mouryan period in 9th century. "The growth of turmeric is almost a religious rite with the Kandhas and it was to improve its culture and output that human sacrifices used to be performed since mid Nineteenth century. It is their chief source of income, as they export it in large qualities, bartering it for grain and salt to drivers of pack bullocks, who come from Ganjam, Sambalpur, Cuttack, Puri and Tributary States (Dist.Gazetter, 1908)."

Table 3.1: Types of crops grown on podu field

Cereals	*Millets*	*Pulses*	*Cash crops*	*Vegetables*	*Oil seeds*
Kudinga (Paddy)	*Arca*	*Masanga* (*Biri*)	Sidiganga (Turmeric)	Sainga (Lab lab)	*Sorasa* (Mustard)
Maize	*Rende*	*Munguka* (*Biri*)	Ginger	*Jadeka* (Banana)	*Rasi* (*Til*)
Jower	*Kuiri*	*Kanga* (*Pegion Pea*)		*Sejeka* (Brinjal)	
	Dedi	*Kaltaka*			
	Klingiraka	*Kadkanga / jhudunga* (*Cow pea*)			

Source: Primary data collected from the field study.

Turmeric thrives well in a sandy or clay loam soil. Such type of soil is available in an area of 1.43 lakh hectares of upland in the district out of which turmeric is grown only in 10,000 hectares.

Production Strategy

The Kandhas prefer podu field or hilly areas to cultivate turmeric. For turmeric production, *balubira* (sandy soil) is preferred because the tuber can grow well in the sand. As the podu land contains sandy and black soil types, the Kandhas generally prefer it for the cultivation. In cultivation of turmeric no fertilisers, pesticides and insecticides are used because the podu field supplies adequate nutritional requirement for the growing plant. As the turmeric has also the quality of germicide, the pests and insects do not affect it.

The entire turmeric production cycle is around one year. After mulching, the burden of work comes down. In turmeric production, maximum labour requirement is called for during the time of twigs collection. Both men and women collect twigs of *sal* trees from the nearby forest. Using a serrated knife they cut down the branches hanging around the trees. The same day or in following days, the Kandhas carry all the branches and cover the turmeric field like a sheath. The land is then left untouched for nine to ten months till the time of harvesting. During harvesting time, both men and women contribute equal labour. While the men dig the soil and collect the tubers carefully, a woman picks up the tubers out of soil and cleans them properly. After the collection of turmeric, women carry it to their home. They use bamboo baskets to carry the turmeric.

Processing

Within a few days all the harvested turmeric are processed. The turmeric is kept in the open field and allowed to dry up. After some days, the Kandhas boil the turmeric in a high temperature. The boiling process is done carefully by the women members. A big hearth is made on the ground. The women plaster around the hearth with mud or sometimes with cow-dung. After two to three days of readying the hearth, the fuel woods are used to provide enough heat for turmeric boiling. The turmeric is kept in a big pot and boiled at high temperature. While boiling, the container is covered with a plate. It is done to retain the steam inside the container. After boiling for three-four hours, turmeric is collected from the pot and brought in the open air. They are allowed to dry completely for a few days. After complete drying, the turmeric size is reduced to one-third of its original size. The cortex of the turmeric becomes deep yellow while the sheaths become greyish brown.

Use of Turmeric

Turmeric is not only used as a spice ingredient to prepare curry but as an important medicine. The Kandhas know that the turmeric has antibiotic and antiseptic properties. It is used as a medicine to cure stomach infection, skin disease, heat effect of the body, etc. After all, the Kandhas use it as a panacea for all diseases. Apart from the medicinal importance, turmeric has ritual importance too. In each and every ritual, turmeric is used as a symbol of purity. The girls after puberty and the women after menstruation consume turmeric. They use the turmeric paste in order to avoid any sorts of bacterial or fungal infection.

Traditional Knowledge of Crop Rotation

In the past, a farmer regenerated the fertility of the farmland by changing or rotating crops, inter-cropping or multiple cropping, and leaving fields fallow. The production practices of traditional crops show their traditional wisdom. Knowledge about soil fertility was well known. As already highlighted, there was practice of intercropping and crop rotation in traditional agricultural practice. Although the practice is very old, yet scientifically it is considered as an advanced technique to improve the soil fertility and facilitate soil conservation. Intercropping constitutes a better use of the land and available rainfall, when the plants compliment each another.

Cultivation of short duration varieties of crops avoids drought situation and ensures safe yielding as compared to long duration varieties. Crops like millets, pulses require less water as compared to paddy. Kandha lands being mainly hilly terrain and upland the kind of millets preferred to be cultivated. For various needs such as fruits, fodder, timber and fuel, trees have to be planted on sloppy uplands as these are not susceptible to drought. The Kandhas grow turmeric and ginger on sloppy lands in order to avoid excessive water that affects the tubers.

CHOICE OF CROPPING PATTERN

Practice of inter-cropping or mixed cropping is better than mono-cropping in the rain-fed areas, which ensures some yield during the aberrant weather conditions or pest problem in stead of complete crop failure apart from providing higher yield during the good years.

The practice of sowing another crop in the standing crop before harvesting reduces the cost of land preparation and water requirement for second crop and thereby avoids drought.

SETTLED CULTIVATION

In traditional agriculture, settled cultivation has also a significant position. But settled cultivation is not as old as the podu cultivation. Before India's independence the settled cultivation did not hold the significant position among the tribals. Podu cultivation provided the basic food requirement and was suitable for them to produce the needed cereals and pulses, the people didn't bother for an alternative immediately. In addition, there was no legal restriction in podu cultivation. The use of simple tools and techniques did not affect their production. Whatever the yielding they had was sufficient for their limited population's requirement. The climate too favoured the practice. The surrounding dense forest was congenial to maintain balanced moisture and temperature of the soil. Based on these, the podu cultivation was prevalent in the region. As the podu cultivation was common in the region, the Kandhas gave less importance to their plain agricultural land. The customary behaviour of the village ritual leader to continue the *'Saru Penu puja'* and uphold the practice was another reason. There was traditional norm to cultivate the podu land as a symbol to respect the mountain mother. As a result, the tribals were guided to continue the practice of mountain cultivation without any hesitation.

The consumption pattern during that time was mainly associated with the crops cultivated on podu land rather than plain land. As the crops grown were easily adaptable to *podu* land rather than plain land, the tribals preferred to continue with podu cultivation. In this way, the consumption habits and climatic adaptability were some of the important factors to adopt podu cultivation as customary practice.

TRADITIONAL LANDHOLDING

Traditionally, the land situated on the mountains was communal property. Private ownership of land could not probably be possible in that period when the shifting cultivation was not easy to carry out alone. The united actions in defence and offence as well as to protect the crops from the raiders and the wild animal did not allow them to go for private ownership

(Pathy, 1984). The Kandhas had no permanent rights except to a piece of land around the site of the village. The evolution of ownership rights among the Kandhas of Kandhaland broadly reflects the predominant land character in the Mahals during the nineteenth century (*ibid*). Since the tenth century the states surrounding the Ghumsar region periodically tried to assert their sway over Kandhaland. The Kandhas resisted outside political dominion till the seventeenth century. This was when the Oriyas began to settle in fortified villages within Kandhaland and cultivated on irrigated land around the settlements. To facilitate administration, the area was divided into *Muthas* (10-15 villages) under the control of the headmen who were supposed to pay a tribute to the raja (*ibid*). However, in due course of time, more particularly, after the independence the village tribal council distributed the podu land to the concerned villages. In 1960s, the village councils again distributed the podu land to the individual households. The lands were distributed to the households based on the family size. In the same way the distribution of plain lands was done with the help of tribal council and village council members. Each household was distributed with the lands of different types. Many landholders still do not have record of rights to their landholdings. Community responsibility and Kui Samaj in the region have managed to avoid any land related conflicts.

The traditional landholding pattern of the Kandhas shows that the land inheritance follows according to patrilineal rule. The land is directly inherited from father to son. A daughter never demands her share on father's property. In feudal system, some households from among the Kandhas were selected in a particular territory to collect the taxes from the landholders. These selected groups were surnamed as Mallick. The Mallick Kandha families were given extra land as reward for their effort to collect taxes from other families and deposited to the *mutha* heads who are mainly the non-tribals from the plain regions. In that way, traditionally, the mallick families in the same community possessed more landholding than their counterparts. The Mallik families after collecting tax from the respective households deposited the same with the *muthadar*.

PRODUCTION OF TRADITIONAL CROPS IN THE SETTLED CULTIVATION

Over the centuries, the Kandhas have also been growing traditional crops on the settled lands. In *saiki siati*, the kitchen garden, the Kandhas grow vegetable plants like papaya, *jadeka* (plantains), pea varieties like *sainga* (Lablab bean), etc.

Sainga (Lablab bean) is a common vegetable crop variety in the region. The crop is preferred due to drought resistance nature. As the crop is of creeper type, the Kandhas use large thorny branches of bamboos and sal branches to support the creeper. The production strategy of the *sainga* is not very complex. The Kandhas simply make furrow on the soil with the help of wooden spade or *pikash* and prepare the soil bed for sowing the seed. The women continue watering plant until the plants grow to a medium size. After a particular growth the creeper is supported by the *ranja* (creeper supporter). It allows the creeper for growing on it. They are also planted near, but inside the fences to support the creeper in order to protect the plant from the cattle.

There are different species of *sainga* being grown in the region. The species are identified based on the pod and seed shapes and also colour. The colour of the pod and the seed varies across species. The Kandhas, at present use seven *sainga* varieties that are of different species but belong to same genus.

The common variety of *sainga* is the medium size pod with both ends curve like U-shape. Each pod varies in size from two to three inches. The seed after getting dry turns black in colour and round in shape. The round seed is usually considered as of good quality. There are also some varieties with flat pods. While the former one is used for perched rice, the latter is used for the curry. The colour of the pod varies from a little greyish green to dark green. Some are a little dark in colour.

The Kandhas select the crop based on the taste too. According to the Kandhas the *sainga* with round shape and green colour pod is often preferred due to its taste. The round-shape *sainga* is the most preferred one because of its yielding and disease resistance qualities. This produces more seeds and less prone to disease.

Jadeka (Plantain)

The plantain, locally called *jadeka*, is one of the oldest vegetable types among the Kandhas in the region. As the region is rich in natural vegetation and the springs flow down across the land, the Kandhas produce the plantain on the banks of the spring. However, there was no importance of *jadeka* cultivation in those days due to less preference among the Kandhas for its consumption. The local varieties of plantain are usually bigger in size and possess rich nutritional content. For a longer time, the *jadeka* attracted a number of traders to the regions because of its size and taste quality. However, *jadeka* could not be a popular crop due to less market opportunity and consumption demand in the region.

Jowar

Jowar is a cereal variety mainly cultivated on the *dhipa jamin* (uplands) during rainy season. The Kandhas prefer sowing of seed on a specific direction of the land position before the onset of rainy season. Mostly, the sowing of seed is done with a gap of 5-6 inches in a longitudinal direction. The logic of leaving the gap in seeding is to allow the growing seedlings to receive enough air and water. The emerging seedlings grow better than the plants having no gap. The old varieties of jowar are healthy and each *kenda* (bunch of pods) produce almost half kilogram of jower. The jowar is too preferred because of its healthy nutritional value. Both the Kandhas and Panas prepare *halwa* and other items out of jowar. The Kandhas boil the seed and produce gruel. They add kanga or other boiled cereals with the jowar gruel to make a little solid type for their consumption. In fact, jower cultivation continues in some part of the district, but gradually losing its ground due to restriction on podu cultivation and domination of paddy cultivation on the settled land. After the introduction of new varieties of paddy and chemical pesticides the soil too was no longer suitable for the production of jower.

Ginger Cultivation

The Kandhas started ginger cultivation way back in the twentieth century. This cultivation began with the interaction of the non-tribal traders with the tribals in the region. The ginger cultivation, in a limited period of time, took an important place in the form of major cash crop production along with the turmeric. Both turmeric and ginger cultivation was done on the hilly land and on the plain land where the soil was usually sandy type. This sandy soil is called as *balubira*. The ginger quality from the region was very good. It produced good aroma. The size was usually bigger and produced good stems. In traditional ginger growing, there was no application of fertiliser as the soil was more fertile. The traditional way of seed treatment was done by allowing the seeds be dried in the sunlight. The seeds are stored on a dry and cool place for the next season's cropping. While the branches of the main tubers were sold, the main cortex was stored for the cultivation purpose. In due course of time, the ginger cultivation has been stopped in the region. There was frequent use of new varieties of seeds imported from the state, Kerala. As a result, there was complete mixture of both traditional seeds and imported seeds in the region. This mixture resulted the loss of original quality of traditional local seeds. The local seed which is better adapted to the local climate was not available easily to the farmers. Again, the imported seeds didn't suit to the regional climate. May be because of these reasons the ginger cultivation stopped in the region.

Mung and Mustard

Mung and mustard seeds are used for mixed cropping. The Kandhas sow the mung seeds, a little before paddy harvesting, i.e, towards the end of rainy season. The same land is used for mustard cultivation a few days after the mung is sown. In Kandha villages women play a major role in sowing seeds. The Kandhas think that the use of both the types in the same field produces a better result.

Paddy Production

Paddy is the most preferred crop among the traditional crops cultivated on the plain land. Traditionally, the Kandhas cultivated more than fifteen varieties of paddy. Some of which, they inherited from their fore fathers and some from the outsiders such as itinerant traders coming from Ganjam, Nayagarh and other regions.

Till 1970 the people used to cultivate only these local varieties in their field. They cultivated those varieties that are suitable to their land, their consumption pattern, ritual behaviour, profitability, etc. The adoption of multiple varieties was done since the time inception. Over the years, it is said that the farmers developed most of the varieties by frequent experimentation and observation.

SELECTION OF CROP AND CROP VARIETY

There was time when a large variety of crops were grown in the region. However, these crops are restricted to millets, cereals and a limited variety of pulses. Among the varieties, millets were preferred by the Kandha as they were suitable for preparing the gruel. So the consumption behaviour was the major factor for the selection of crops in traditional agriculture. The other factor was the climate of the region. In podu cultivation the crops grow well. According to the Kandhas respondents, they abide by the rules of the nature. They would prefer to continue with that which is well adapted to the local climate and culture. People preferred the millets such as arca, klingiraka, dedi, etc as these crops are well adapted to the climate of that time.

RATIONALE IN SELECTING A VARIETY

The better yielding of paddy is not always preferred. Their consumption pattern determines the preference. For example, traditionally, the Kandhas prefer the varieties, which produced better gruel. There are many other factors, which also determine the selection of the variety. For example, the Kandhas also consider the resistance qualities of the crop, the water and temperature requirement of the varieties of the paddy, etc. Some

Table 3.2: Local paddy varieties

Sl. No.	*Name of the variety*	*Characteristic of the variety*	*Use pattern*
1.	Mata Saka	Dwarf and fat, white colour, produce better gruel	Daily
2.	Nadia satanga	White, thin and medium size, scent normal, yield medium, plain land can be used for the cultivation. More grain	Daily
3.	Kalitolashaka	Medium, thin, black colour scented, yield capacity-less. Low land will be preferable	Occasional, during festival and rituals. Used as *khiri* item
4	Bukuranga	Very thin, smaller size, white colour, strongly scented, can be grown on uplands. Yield-medium. Early variety	Occasionally used during rituals and festivals
5	Ajalanga	Big size, white, yield medium, upland variety	Daily
6.	Brihadlanga	Very fat, long and big size. Red colour. Lowland is better. Resistant to pest and insect infection.	Daily
7.	Kundadhan	Big size with big crown, red colour, better husk production, upland is suitable. Early production.	Daily
8.	Tureka	Medium size. In any type of land Early variety	Occasional, mainly for preparing *lia* (a food item) and pressed rice
9.	Nagar dhan	Medium, red, low land, need much water, late variety	Daily
10.	Daspalianga	Red colour, big size, low land is preferable	On special occasion for preparing *lia* and *mudhi* (puffed rice)
11.	Putkudianga	Small, scented, low land variety	Occasional use

varieties are preferred for having ritual importance. For example, *kalitolashaka* and *Bukuranga* are preferred because they are strongly scented and better for *khiri* preparation. The Kandhas choose certain varieties even when the yielding of these varieties is less. *Tureka* and *daspalianga* are preferred for preparation of puffed rice and *lia*.

Similarly, the selection of *sainga* is based on the criteria like suitable climatic condition, more yield and easy production and consumption preference. *Sainga* is also preferred for its taste and nutritional qualities. Apart from the *kanga*, *sainga* is suitable to be used as an ingredient for the preparation of gruel by adding it with rice.

TRADITIONAL KNOWLEDGE OF CROP DIVERSIFICATION AND OTHERS

The cultivation of multiple varieties of paddy depicts the nature of diversification in agriculture. The adoption of multiple varieties, which is there since centuries, is the result of their experience. Historically, their ancestors have developed these varieties through their knowledge and experiences. Today their products are the result of multiple experiments already done by their predecessors. The Kandhas have superior knowledge to select the varieties through frequent scrutinising.

The scientists in agricultural research centres develop hybrid seeds, produce genetically modified crops by selecting the pollen of new and superior plant species through the process of transfusion, and develop new mechanisms for better yielding. In tribal communities, the knowledge in cross-pollination or genetic modification may not be there. But the tribals do it differently. The Kandha farmers scrutinise and select the best seeds out of the existing varieties. They store them and cross check them in different soil and different climatic condition. They observe which varieties grow well in which type of soil and under what condition as in the case of paddy selection. In entire Katingia panchayat, the Kandhas have generated improved varieties of Turmeric, French beans, and *Lablab beans*. More particularly, they strongly claim that the improved local bean varieties are their local property. Some people are aware of patents and patent rights. They claim that they need to get patent for this local variety. They have developed this variety by their cumulative effort. Regarding this Harekrishna Pradhan of Sudhipara village said that the selection of pure seeds for further cropping needs an experienced person who is already in this field for over the years. We select some pure seeds based on its size, shape, colour, smell and others. We protect it from heat and store in a normal temperature. The dry seed is stored in dry places only. All farmers do this. But a few get only good seeds and yielding as well. So the seed out of selected pods is again kept safely and people observe and experiment as to what possible extent they can collect good seed for further sowing. So, people after experiments in different climatic condition reach to the conclusion as to which one would yield more under the particular circumstances.

The better yielding varieties are used for the daily consumption and less yielding varieties for occasional and auspicious events. The Kandhas say that the varieties, which yield more, are more prone to diseases than the less yielding varieties. The less yielding varieties usually possess some better qualities like odour, taste, appearance and disease resistance.

Though the belief system among the Kandha was so strong and it influenced agricultural practices, innovation was never lacking. According to Choudhury Pradhan, a 107-years-old man:

> 'The people had strong belief on nature and supernatural objects. Frequent experimentation on agricultural practices continued to get better production out of the podu land. However, any loss to crop was often considered as the cause of the wrath of these supernatural powers. The production of multiple varieties of millets is the output of their traditional wisdom. He claims that the millets have never been borrowed from other communities. In due course of time the people developed these varieties along with some of the crops that already existed. People developed and adopted technologies based on the needs of the individuals and the community. Unnecessary exploitation of nature was against the norm of the community'.

INDIGENOUS KNOWLEDGE ABOUT CROP PRTECTION

The pest control mechanisms among the farmers included the using of traditional herbs and other ingredients. In paddy fields the Kandha farmers have been using sal sticks or *salap* leaves (sago palm leaves) to check out the pests and insects. They fix these leaves and keep them in the centre of the paddy field for not allowing the insects or pest to grow over there. The farmers say that the salap and sal leafs have some scents and this keep the insects at distance from it. This also has some chemicals for which the insect never attack this plant. Therefore, the use of the stick and leaves of these plants becomes helpful in preventing the insect and pest infection to the growing crops. Sometimes, they collect spiders from the old houses and leave them in the paddy field. This shows the indigenous technical knowledge of the Kandha farmers.

WATER MANAGEMENT

The water management mechanism in traditional agricultural system in tribal areas is the result of interlink between social organisation and availability of resources. Natural flow of water in the form of small spring and springs is common as the area is deeply surrounded by the mountains and dense forest. In Katingia the water flows down to the plain areas from the fringes of the hills. Traditionally, the Kandhas utilised this for drinking as well as for the agriculture purpose. The spring water never goes waste, ever since the Kandhas started practising settled agriculture.

TOOLS USING IN TRADITIONAL AGRICULTURE

The Kandhas had specific social beliefs in using traditional tools for the *podu* cultivation. Many Centuries back, the Kandhas never cut the crops with a sickle having serrated edge, such as used by the plain Oriyas. They used straight-edged knives for the above purpose. Cattle were not used to thresh crops after they were harvested. The Kandha used to do all the works starting from threshing to harvesting crops with the help of simple and traditional tools made up of wood and iron. The serrated sickle was not used, which produces a sound like that a cattle grazing, which would be unpropitious. If cattle were used in threshing the crop, it is believed that the Earth God would feel insulted by the dung and urine of the animal (Thurston, 1902). Thus *Podu* cultivation appeared to be not only an economic pursuit of Kandha community, but it accounted for their social life. Group cooperation in labour contribution was essential for shifting cultivation. This group activity with specific norms had cherished some values. As already discussed the podu land distribution was managed by the village committee and the sharing of labour was maintained by both men and women. There was no conflict over possession of land and sharing of produce. In this way their social structure, economy, political organisation and religion were all accountable to the practice of shifting cultivation.

The major characteristic of the traditional agriculture in the region is the use of only traditional tools and handmade local tools. The oldest form of ploughing land is by cow or bullock. In addition, traditional tools like *pikash, rampa,* etc were used for the purpose.

OTHER TRADITIONAL TOOLS

The cultivation of traditional crops was also associated with the use of traditional tools and techniques. The use of traditional tools and techniques varies from simple to complex ones. The following tools are used for the traditional crops in the *podu chas*.

Kodi	–	Preparation of land
Kata	–	For making *siara* and canals
Pikash	–	For ploughing hard soil.
Gadi	–	For digging soil
Langal	–	For ploughing
Kuradhi	–	For cutting trees
Korada	–	To prepare plain land in hilly area

RATIONALITY IN ADOPTION AND USE OF TECHNOLOGIES IN TRADITIONAL AGRICULTURE

Most of the technologies in the tribal areas are derived from nature. Their knowledge is based on making the maximum use of the forest. When the market opportunity was less followed by the inadequate support from the public and private agencies, the farmers had to depend on the traditional tools and technologies for the agricultural purpose. All the necessary implements were derived not only from the forest, and from the other available resources also. *Pikash* is the traditional implement developed to break the hard soil. This was developed based on the climatic condition of the region. *Korada* is the tool used for pushing the soil from the uneven land in order to make it plain. The tools developed were less energy consuming and easy to handle. Use of manual labour was enough to handle the tools. The techniques to handle these have been inherited from their forefathers. This process continues and transmitted to the next generation. Moreover, leaving tools and techniques adopted by the forefathers is against the morality, as the Kandhas said.

In traditional agriculture, the adoption and preferences among the Kandha is mainly on ecological base. The introduction of new technologies was not possible as the area was having little communication facilities. The market facility was beyond their expectation. Communication with the outside world was very limited. That did not allow them to access to new opportunities. They had little communication facilities, which made them to be restricted within.

CONCLUSION

The above study on tribal agriculture reflects a few points on the traditional beliefs, knowledge and practices in agriculture and focuses the uniqueness of the community. Their beliefs in origin of agriculture, prediction of climate through birds, beliefs in nature and supernatural entities are important part of the community culture. Shifting cultivation is the oldest and most traditional form of agriculture. The process of shifting cultivation, crop production activities, crop rotation procedure, water management, and management of retention of soil moisture signifies the knowledge of Kandha community. Millets like *kuiri, arca, klingiraka,* etc were most preferred food items while jower, maize and paddy are the main traditional cereals. The Kandhas have identified different varies of millets, pulses and cereals. Even they have identified multiple varieties paddy. The selection of crop is based on the basic principles and rationalities. They have developed their own agricultural tools and implements, water management mechanism and crop protection techniques.

❑ ❑ ❑ ❑ ❑

CHAPTER - IV

TRENDS IN AGRICULTURE AND ADOPTION OF TECHNOLOGY

The post-independence period in India has envisaged significant changes in agriculture both in rural and tribal areas. The tribal areas are lagging behind with regard to taking up new agricultural practices and technological adoption. However, tribal economic and agricultural practices continue to be changing towards modernisation over the years. Vidyarthi and Upadhyay (1980) have delineated the changing economy of the Kharia. Many a times, the Kharias have taken advantage of market economy. Dudh Kharias have shifted to commercial agricultural production including cash crops. Similarly, among the Korwas of Madhya Pradesh and Uttar Pradesh, there is a shift from traditional shifting cultivation to plough cultivation (Sandhawar, 1990, 115-218). Singh (1982) points out that modernized agricultural farming is emerging among the Santhal and Oraon of Chhotanagpur, the Gond and Korku of middle India, and the Badaga and Mullu Kurumba of the Nilgiris. Cash cropping and coconut production has turned Nicoberese into an affluent community. There are many other instances where tribal and rural peasants have adopted modern agricultural practices.

In some parts of tribal pockets in Orissa, horticulture and other commercial crops are well accepted. For example, the Dongria Kandha in Rayagada and Kandhamal districts are involved in horticulture (Aparajita, 1994). Vegetable cultivation is widely accepted among many tribes such as the Bhuyan, the Munda, the Santhal, the Kandha, etc. In Kandhamal district, of the total area, 1,10,470 ha was covered by field crops during 2002-03, paddy was covered in 40% area followed by 9% with pulses, 11% with oilseeds, 12% spices, 13% vegetable and 15% with other cereal crops like maize, ragi, wheat and other minor millets(XIM, Bhubaneswar, 2006).

AGRICULTURAL PRACTICES IN THE VILLAGES

With continuation of traditional agricultural practices in the villages, there are also modern agricultural practices observed in the present form of agriculture. There is adoption of vegetable crops, spices, and cereals by the people. There is change in adoption of technology. People have embraced new mechanized inputs in agriculture. The details about the recent agricultural practices in the villages are discussed in the later part.

LANDHOLDING

Land is prerequisite to agriculture. The quality of soil, position or location, and size of land determines the crop production. Therefore, the discussion about landholding is very significant before going into detailed about agriculture.

G. Udayagiri block in Kandhamal district has total geographical area 25102 ha. Of which forest area alone covers 12801 ha and miscellaneous trees and groves 146 ha, land put to non agricultural use is 584 ha, and total cultivable area is 9658 ha. Of the total cultivable land, the distribution of high, medium and low land constitutes 7858 ha, 800 ha and 1000 ha respectively. In the district total irrigated area is 25462 ha and total cultivable irrigated land in G. Udayagiri block is 2236 ha[1]. This shows that less than 25% of cultivated area in G. Udayagiri is irrigated. As far as land distribution in the district is concerned, more than 80% of total land holders in the district are small and marginal land holders.

Coming back to the village study, it is estimated that the total cultivable land of the people in Sudhipara and Laburi villages is approx. 220 acres and around 82 (90.1%) Kandha households of the total 91 Kandha households hold land of different sizes. While 55 (60.4%) Kandha households have both irrigated and non-irrigated land, 7 (7.7%) households have only irrigated land and 20 households have only non-irrigated land. Among Kandha-households 9 (9.9%) have no land. Thus, 62 (68.1%) households have irrigated land among the total 91 Kandha households. Irrigation is mainly supported by small canal and natural water flows of springs flowing around the village.

As far as landholding among the Pana is concerned, it is observed that 40% households have no land at all. There are only 13.3% households having both irrigated and non-irrigated land.

Majority of the households in the villages have marginal landholdings, which comes under 0.5 to 1.0 acre. The landholding details of the villages show that 84 households out of 123 households have inherited land of different sizes. Twenty-two Kandha households and 15 Pana households have not inherited any land. It is evident that the landholding by inheritance among Pana is not common, but among Kandhas this is definitely one of the issues. Most of the households who did not inherit their land are mainly due to selling away of land by their forefathers, which is discussed in subsequent paragraphs.

As far as land purchase is concerned, only 14 Kandha households have purchased land during last four decades. The transfer or selling of land is mainly taken place among the Kandhas. A Kandha sells land when he needs money urgently to bear the expenditure of daughter's marriage, death ceremony, and others. In the case of sudden crop loss there are also some people who need to sell their land. In Laburi and Sudhipara villages three families have sold land to farmers from Katingia village. Similarly, Sudhipara and Laburi villagers have purchased land from Budedpara, Lingagada, Raikia and from Bradinaju villages. Exchange of lands takes place among the tribals. But it is the rich farmers who take the advantage. In Sudhipara and Laburi villages the families who have purchased land are mainly rich farmers and having more than Rs 80,000 annual income from agriculture. Five decades back F.G. Baily traced out several reasons about how the land came into market from Bisipara village study in Phulbani district. Some of his observations are: to pay for mortuary rites, to bring bride, to send a bride, to buy plough cattle, to repair land damage, to buy food, to fight a case, due to loss in trade, for investment in cultivation, etc. However, even after fifty years of Bailey's study the reasons for selling land are almost same in many villages of Kandhamal district. In Sudhipara and Laburi villages also these are the reasons for selling away the land. According to Ramachandra Mallick, a farmer of Sudhipara village, "a landholder does not sell his land until he comes across extreme pressure. The reasons of selling land are: when the daughter gets married, to buy plough cattle, or when there is no caretaker in the family to undertake agricultural activities and in such cases when the lands do not give good returns. But lease is also an option for the people who want to get some money urgently."

Table 4.1: Ownership of Land among Panas and Kandhas (household-wise)

Types of land	*Landholding by each social group (Number of Households)*			*Total number of households*
	Kandha	*Pana*	*Other*	
Only Irrigated Land	7 (7.7%)	0 (0%)	0 (0%)	7 (5.7%)
Only non-irrigated Land	20 (22.0%)	14 (46.7%)	0 (0%)	34 (27.6%)
Both Irrigated and non-irrigated land	55 (60.4%)	4 (13.3%)	0 (0%)	59 (48.0%)
No landholding	9 (9.9%)	12 (40.0%)	2 (100%)	23 (18.7%)
Total	**91 (100%)**	**30 (100%)**	**2 (100%)**	**123 (100%)**

Table 4.2: Land-holding details

(a) Inherited land by each social group

Inherited land	*Group-wise distribution of landholdings (No. of households)*			
In acres	*Kandha*	*Pana*	*Other*	*Total*
0-0.5	42	11	0	53
0.5-1	15	4	0	19
1-1.5	11	0	0	11
>2	1	0	0	1
Nil	22	15	2	39
Total	**91**	**30**	**02**	**123**

(b) Purchased land by each social group

Purchased land	*Group-wise distribution of landholdings (No. of households)*			
	Kandha	*Pana*	*Other*	*Total*
0.0-0.5	11	0	0	11
0.5-1.0	2	0	0	2
1.0-1.5	1	0	0	1
Nil	77	30	2	109
Total	**91**	**30**	**2**	**123**

In addition to the above reasons, there are a few additional reasons like pressure from banks, LAMPCS and other cooperative societies from where the Kandhas have borrowed money. According to a few marginal landholders in the village, the pressure from LAMPCs officials is more than the SBI officials. They display the name of defaulters on the blackboard in front of the office. They come to village and give warnings. In case of failure, they lock the house. So a borrower has to try soon so as to repay back his loan. In case he is not able to arrange money from other source he has to sell or mortgage his land. In these two villages, Laburi and Sudhipara, there are two cases where the lands have been sold for repaying loan.

Some families with lack of social and economic security want to sell away their land; however, preference is to sell to the members of relatives. With the absence of relatives who can buy land, one looks for one's friends or relatives' friends for selling land. This is mainly because of obligation to relatives and friends within the same community and for some it is the restriction not to sell land to non-tribal due to scheduled area land transfer act, 1956. There is no instance of selling land to non-tribals in the villages.

Similarly, among Pana households there is no instance of purchasing land. From this, it is confirmed that the selling of land to non-tribals has been stopped in the villages and there is no instance of illegal land transfer in present days. Besides land transfer regulation act, increase in literacy in the villages may be one of the important factors to restrict transfer of land to non-tribals. Many people understand how important the land is. So, selling land is not a solution to deal with the economic problems of family.

The Kandhas prefer sharing or keeping land in mortgage to selling away of the land. Among tribals, the number of households possessing land under share cropping and under mortgage are 11 and 23 respectively. The observation in the village is that marginal landholders who are economically insecure and have not enough family members to work in the agricultural field usually lease out their land. Similarly, Kandha families of other villages who have land in Sudhipara villages preferably lease out their land to the Kandha families in Sudhipara and Laburi villages.

The total numbers of Pana households having landholding under share cropping and under mortgage are only 2 and 3 respectively (Table 4.3).

Table 4.3: Land under sharecropping and under mortgage

(a) Land under sharecropping

Under sharecropping In acres	*Social groups (No. of households)*			
	Kandha	*Pana*	*Other*	*Total*
0-0.5	10	1	2	13
0.5-1	1	0	0	1
Nil	80	28	0	108
Total	**91**	**30**	**2**	**123**

(b) Land under mortgage

Under Mortgage	*Kandha/Pana/Other (No. of households)*			
	Kandha	*Pana*	*Other*	*Total*
0-0.5	68	27	2	97
0-0.5	19	1	0	20
0.5-1	4	1	0	5
1-1.5	0	1	0	1
Total	**91**	**30**	**2**	**123**

The land reform policies of Orissa have more or less protected the tribals. But concealed land alienation is taking place in some other parts of tribal areas.

DIVISION OF LABOR IN AGRICULTURE

Labour is primarily confined to household members. All the members of the family across age and gender contribute their labour in agriculture. The division of labour is mainly maintained on the basis of age rather than of gender. The young members work for low labor input agricultural works such as weeding, carrying food from home to field for their elders and so on. Sometimes, they are engaged for collection of timber and wooden poles from the forest for fencing the kitchen garden, for accessories of agricultural implements and for fuel. The men engage themselves in ploughing, fencing, harvesting and other activities. The men prepare the field with their manual labour and hard work. The old women stay at home and take care of small kids when the mother is out of home.

ROLE OF WOMEN IN AGRICULTURE

Again, among others, the women play important role in agriculture. The women, irrespective of caste and tribe, are the efficient workers and skilled laborers. Starting from sowing to weeding, and harvesting to threshing the women play crucial role. Right from sunrise or before that the women start their journey from home to agricultural field. They engage themselves whole heartedly all-round the day in the agricultural field during khariff and rabi seasons. A woman is more conscious about taking care of the crops and other agricultural operations. She doesn't take rest until and unless the crops are harvested and brought to the home. However, the responsibility of a woman doesn't end here. Apart from agricultural operations, a woman is the sole active person in family activities too. She cooks and serves all her family members during the early morning and evening before and after the agricultural work. In general, a woman works for more than 12 hours a day for her family purpose.

VARIATION IN HANDLING AGRICULTURAL TECHNOLOGY

While a man handles all the heavy mechanical instruments such as plowing machine, *gainchi*, trencher, heavy-axe, thresher, etc, the women use small and light instruments for agricultural operations. *Gadi*, for example, is a small instrument mainly used for weeding and digging of soil during turmeric harvesting. This type of instrument is commonly operated by the women members. Women also use axe for cutting wood and poles for fencing the garden. Spade is used by the women for digging purpose. The size of the instruments used by women is often small in size than the instruments used by male members.

LOCAL LAND CLASSIFICATION AND THE CROP GROWN IN THE REGION

The local classification of land is based on location of the land and water holding capacity of soil. In addition, this classification is also based on soil quality. The people have classified the land mainly into four types, such as *panga, badanga, taki and dekchedi*. *Panga* land is mainly used for turmeric and *kunda dhan* growing. Though the land has potential for the production of ragi and groundnut, but ragi and groundnut have not been popular like turmeric and *kunda dhan*. This land is less capable to hold water. Similarly, *badanga* land is preferred for French beans and Lablab beans cultivation. *Badanga* is the local term of 'kitchen garden'. *Taki* is mainly plain lands and located in low-lying areas in comparision to *badanga*. This is capable to hold water. Therefore, it is mainly used for paddy and vegetable growing. *Dekchedi* is low land, which is preferably used for paddy cultivation.

Ketanga (land)	**Crops/Vegetables grown**
Panga --------	Turmeric, *Kunda dhan*, *Ragi*, Groundnut
Badanga ------	French beans, *Lablab beans*, Maize, Mustard
Taki ----------	Paddy and vegetables
Dekchedi -----	Paddy

Other local terms used for identification of land is based on the location and water holding capacity are as follows

Local term	**English term**
Kuiketang ----------------	Dry upland
Siruketang ---------------	Low land
Pangamaha --------------	Upland with irrigation support through water reservoir
Dakuri -------------------	Land below water reservoir
Sudisuga -----------------	Low land near *chuas*
Jadiganda ----------------	Land near springs
Sahinera ------------------	Land near lower end hamlets
Madiketanga -----------	Medium plain land

The above local terms are mainly developed by people of the village to identify their land. In addition, different terms are also based on quality of lands. People evaluate the quality of land based on the local terms.

TREND OF VEGETABLE CULTIVATION IN THE REGION

In 1970s the vegetable cultivation was introduced by a Kandha person, Gapal Pradhan, of Sudhipara village. He collected some seeds of the French beans from Phiringia region when he was working as a school teacher in that region. He asked some of the agricultural officials and the local people who had already taken up vegetable cultivation about the procedure and strategies of vegetable cultivation. After acquiring the skills from the people and the officials, he sowed some seeds following specified procedures. This became a successful endeavor for the willing person to take up the activity further. He collected other varieties of seeds and fertilizers from the cooperative agencies and adopted the practice rigorously. His innovative thoughts, knowledge, interest and eagerness, and awareness made him a successful farmer in the region. In due course this knowledge and the practice diffused and reached to other persons having similar interest. People adopted vegetable cultivation after the success came in the form of good yielding of vegetables at the innovator's end. Gopal Pradhan is no more alive but he is remembered for introducing vegetable cultivation in the Katingia panchayat. According to many people, 'Gopal Pradhan' became a transmitter of knowledge regarding use of fertilizer, pesticides and insecticides, and other technical procedures in agriculture in the region. He could learn about fertiliser application on crop from the VAWs coming to villages to create awareness about fertiliser application on crops and use of HYV seeds.

Now, in the region, many people have taken up vegetable cultivation. More than seventy percent of the total households in the villages have more or less vegetable cultivation of which the beans cultivation is prominent in the region. Other vegetable crops such as cabbage, cauliflower, *lablab beans*, and potato are also cultivated in the villages. The knowledge of vegetable cultivation has spread to other villages in the region. Some of the people of the village claim that their daughters after getting married have enabled the members of in-laws to get into vegetable cultivation. These daughters from the villages share their experience with in-laws and help in using certain procedures like use of fertilizer, pesticides, insecticides, and others in vegetable cultivation.

The successful adoption of the beans cultivation is because of the repeated efforts by farmers. A few successful farmers pointed out that they were not successful to get good returns from the crop initially. In latter period they used *khat* (bio-compost) for the plants to grow better. They could learn to cut the branches in the middle of its growth for better growing of the plant. This idea could be developed from interaction with the VAW and agricultural extension officer in the villages.

In the last thirty years, the plants have been grown in different soils, under different types of lands, different weather condition, etc. They have observed both success and failure under different geographical situations. They have used different types of seeds and experimented to see which seeds result in better growth of plants and ultimately produce more, and give better aroma and better taste. They have also used both compost and chemical fertiliser and observed which is suitable and affordable. In this way they have experience of crops growing under different situations and conditions also. It is observed that the people prefer kitchen garden for the beans cultivation, which is easier to guard from cattle, from birds and other anti-social agents that spoil the plants. Having situated near houses and occupying a small area, it is also easier and risk free applying compost and bio-fertiliser on crops.

The case of turmeric growing with different soil types has already been discussed. They have used different varieties of seeds, fertiliser and pesticides as well. However, at last it could be realized that the local varieties in the sandy loam soil produce better yield.

OBSERVATION AS THE KEY TO ADOPTION

Not all the farmers test with different geographical and climatic conditions and follow trial and error methods. There are a few or a very few who do this. However, the people want to know more after observations of a successful practice. The adoption of crops is based on the observation of successful practice. Be it crop or tools if it gives good result people develop interest to take up it. The same was the case with vegetable cultivation. People could observe that the vegetable cultivation was successful in the region after it was introduced by 'Gopal Pradhan'. So the observation of the people from a successful farmer enhanced their interest to come up with such practice and pass it on to further generation.

(Success)

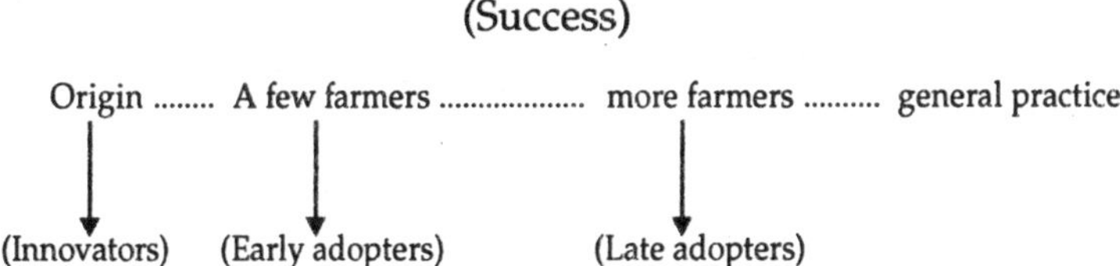

FRENCH BEANS CULTIVATION: PROCEDURES AND PRACTICES

French beans is cultivated twice a year. The cycle of each cropping is 5-6 months. First cropping begins in the month of *Vaisakh* (Apr-May) when the land is made ready after ploughing and application of *khat* (bio-compost). After thorough ploughing, the soil bed is prepared by clearing it of stones and boulders with the help of plough or other local tools. The soil is raised to a little high in order to protect the seed from washing away by water in rainy season. The seeds are sown by hand. In different rows or columns, the bed is raised for seedling. With the onset of rainy seasons, the seeds start germinating. After one month (May-June) the creeper is supported with *ranja*, the branches of tree. In order to have better width of the plant the stem is cut at middle of its growth. As a result, more branches come out of the plant and grow widely around the supporter. The plant starts flowering in *Sravan* (July-Aug) and gets ready with pods in early August. Both

Bhadra (August-September) and *Aswin* (September-Oct) are the yielding season when the farmers collect the produce for selling. Similarly, the second cropping starts in *Kartik* (October-Nov) and ends in *Chaitra* (Feb-March).

USE OF TOOLS IN THE BEANS CULTIVATION

The tools used for the beans cultivation are very simple and traditional ones. Plough made up of wood or iron is used for ploughing. Cows, bullocks or buffaloes are engaged for the ploughing purpose. The families, who do not have these animals, opt for their own labour to prepare soil land for sowing or sometimes hire animals from the other families. With the help of spade/*kodi and korada,* the soil is raised for preparing bed. Some people make round and some make the rectangle beds of different sizes. *Pikash,* a traditional tool, is meanwhile used to make the soil more porous and smooth by breaking the mounds and hard soil.

Over the years, the beans cultivation has brought name and fame to the farmers of the region. People from other regions like Daringibadi, Bhanjanagar, and Phulbani also know this place for famous beans cultivation. The beans cultivation is a family practice and both girls and boys are engaged in cultivation. The girls of these villages are preferred to take as bride by other villages in the district. According to Raghunath Pradhan, a 65 years old farmer, his daughter got married in Gresingia village of G. Udayagiri block. After a few months of her marriage she started persuading her husband to cultivate the beans. However, her husband was initially reluctant because he had not known about the cultivation procedures. For her husband's thinking it was difficult, not because he had no land or labour power, but the practice was unknown to him. However, this girl after marriage changed the thinking of her husband. She became the first to bring such cultivation in the entire village. Over the years the beans cultivation has become a popular practice in entire G. Udayagiri block. But Katingia panchayat has been a successful village to produce more and supply a sizable quantity to the market. Almost all landholders in Sudhipara village and more than 60 percent landholders in Laburi village cultivate the beans. The average land engaged for the beans cultivation is nearly 0.20 acres per households and average yielding per *mana* (0.25 acre) is 3 quintals per year.

The beans production has also attracted private businessmen into the region. Irrespective of small and medium landholdings of the people, the practice has been adopted by large number of households in the region. Growing demand of the beans as a rich nutritional content staple crop as well as cash crop with high market values has more or less encouraged the farmers to continue the practice of the beans cultivation.

In recent years, there is penetration of hybrid seeds of the beans to the region. Despite high demand for the local varieties of the beans, there are some threats from the external agents, particularly, from the private businessmen who sell duplicate seeds in the name of hybrid and improved seeds to the farmers. These seeds look alike the local seeds. As a result, the innocent farmers get trapped with these businessmen and buy these varieties. These seeds collected from the shops yield less as the size of the pods does not go beyond 3-4 inches. However, the local varieties are highly yieldable; the size grows above 5 inches. Besides, the taste of the local varieties attracts the farmers to continue with this. Subsequently, the demand for the local varieties is very remarkable in the region. However, spurious seeds that are available in the local market are a major issue. Sincere efforts need to be made to protect the farmers from the exploitative behavior of private businessmen lest there should be lose of valuable crop from the region.

The Reasons for Success of the Beans Cultivation in the Region

The beans cultivation became very successful due to the farmers sincere effort. It needs timely supply of water, manure, and ranja (creeper supporter) for growing. The farmers are very systematic in doing all this in time. The local climate is favourable for the beans growing. People collect branches of sal trees or other selected trees to use as creeper supporter. Availability of these trees in plenty in the region has also been helpful to grow these varieties.

All the family members contribute labour in different forms like sowing, weeding, watering, fencing the beans garden, etc. Inter family cooperation is also equally important. One family helps other in protecting crops from cattle, allowing water flow to others, and at the time of requirement each family supports other by contributing labour at predetermined wage rate. People charge Rs 40 per day as wage. This price has been fixed by the community for the past three years.

French bean is one of the crops which enable fetching instant profit from a small landholding. That is very important reason why many take up this cultivation. This crop has enabled some people to gain good market price. During the time of field study the market value for the beans was Rs. 9-10 per kilogram. It was observed that the income from one *mana* ranged between Rs. 2200-3000. In addition, this has become a common vegetable for household consumption in recent years. All the households who grow this crop preferably use this vegetable for household consumption. In this way, this crop has become an important cash crop in the region.

In addition to the above factors, the other important factor is possessiveness. Having certain recognition and encouragement from people from nearby regions there is further increase in effort how to sustain the practice. In order to do this, the people are more conscious about retaining soil fertility for further generations. They preferably use manures produced out of cow dung and other organic extractions. Chemical fertilizer is less preferred but a few people use due to unavailability of sufficient compost and other organic manure.

As people are more or less acquainted with the market price, this has aroused the interest among the farmers to be successful adopters of the cash crop. But it is not true to say that there is no market problem in the region. There are also many problems which hinder the enthusiasm of many farmers. These problems are explained in latter part.

CULTIVATION OF *SAINGA (LABLAB BEAN)* AND OTHER VEGETABLES

Lablab bean is a traditional crop of legume variety, still retains its importance mainly due to its consumption value. It is locally called as *sainga*. This legume variety has been cultivated over centuries in the region. This is also their important staple crop much before the French beans cultivation. This is an annual crop. Most of the people including Kandhas and Panas grow this crop at the onset of winter season. Within two months the creeper starts flowering and within one month bunches of pods spring up from many nodes of the branches. This is the most common legume/bean variety in the region. The practice has been based on pure traditional technological know-how.

Production Procedure of *Sainga* (Lablab bean) Cultivation

The production procedure is similar to French beans cultivation. As both of these are creeper varieties, the use of supporter is must. But the only difference is that the sim does not need much attention and fear of insect and pest is very less due to its disease resistant capability. The yield from one plant is more than a beans plant. It has potential to sustain in the eastern ghat agro-climatic zone. Though the production procedure is almost same with the beans production, this is mainly grown in kitchen garden. It also does not require much attention. People are aware of the fact that it will be high yielding if the plant is given as much care as being given for the beans cultivation. But there is no such care taken for *sainga* cultivation.

According to some people the growing of *sainga* also enhances the soil quality. Some educated people in the village are aware of nitrogen fixation of soil from growing of *sainga* plant. It does not require much labour, much technical skill, care, and supply of any other inputs. In sum, this is a low input and high output crop. Therefore, this is favourably accepted by many people in the region.

OTHER CROPS

Other crops are cabbage, cauliflower, and peanuts mainly taken up by the people in the region. These are relatively new in the region.

These vegetable crops are cultivated with the same procedures as followed by the non-tribals in the locality. The use of hybrid seeds is common in the vegetable cultivation. Application of fertiliser and pesticides is increasing with the rise of cultivation of vegetable crops.

MEAN VEGETABLE PRODUCTION (IN QUINTAL) IN THE STUDY VILLAGES

Around 28% (34) households of the villages grow vegetables like cabbage, cauliflower, potato, green pea and others in their field. As shown in the adjacent diagram, the mean productions of the vegetable varieties vary within 6-7 quintals per year per household among the vegetable growers. The production of these vegetable varieties is primarily meant to supply to local and regional market.

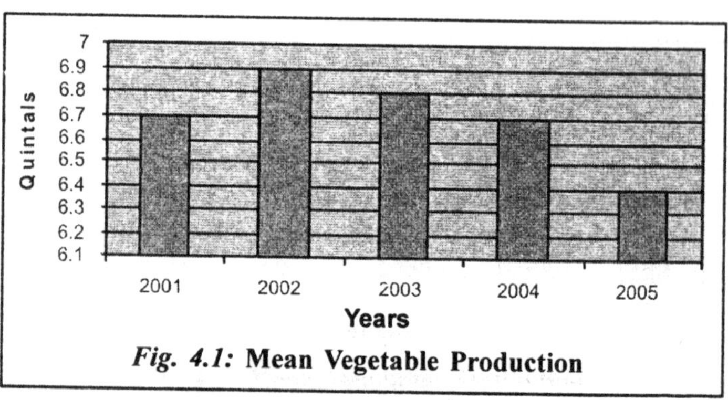

Fig. 4.1: **Mean Vegetable Production**

According to a research report by RRTTS in G. Udayagiri block, the region has potential for horticulture and vegetable production. However, still this has not become possible for many families and many villages in the region due to following reasons:

- **Due to Lack of Financial Support**: Growing of vegetable crops require financial support for application of fertiliser, labour input, pesticides and insecticides, etc. Almost all the people in the villages are aware of chemical fertiliser and pesticides. The vegetables crops are mainly prone to pest and insect infection. As there are no enough local mechanisms to protect plants from insect and pest infection, and to supply the required chemical pesticides. It is not possible for all landholders in the villages to afford them.

- **Fear of Crop Failure**: Some people in the villages have fear of crop loss. As vegetable cultivation needs investment, the farmers expect good return from the crops. But there are instances in the villages in the past regarding crop loss due to pest attack or unpredicted rain fall. During last five decades, several farmers claim that they have experienced crop failure several times. Their opinion is that the crop loss is due to scarcity of rainfall and also due to unpredicted heavy rainfall. The unpredicted heavy rainfall causes more problems to the farmers. As the region is sloppy, the entire crops are easily washed away due to heavy rainfall. So generating profit becomes abysmally low or even zero.

- **No Timely Supply of Input**: Timely application of inputs like water, fertiliser, insecticide and pesticide on one hand and agricultural operations like weeding, raising beds, cutting and others on the other hand are not possible for all the farmers in the villages. This also hinders the interest of some farmers

- **Preoccupation with Other Crops Production:** As already discussed most people grow the beans and turmeric, simultaneously, taking up vegetable cultivation is not easy for many farmers in the villages.

- **Lack of Irrigation or Water Supply**: Not all the households in the villages are having irrigation support or sources of water supply. It is true that more than 60% households have source of spring water irrigation. But during summer season the lands become dry. Therefore, it is not possible for many farmers in the village to adopt vegetable cultivation.

CAULIFLOWER PRODUCTION IN THE VILLAGES

Cauliflower when introduced was initially not successful in the villages. But agricultural officials started giving training for successful operation. Initially, the Kandhas did not know the use of pesticide and insecticide. As a result there was less chance of getting good return from the crop. Similarly, the application of chemicals to keep vegetable fresh was unaware. This too was a fear of the people. However, they could be aware of about the application of the chemicals for keeping vegetable fresh from interaction with the agricultural officials and extension workers. They could avail it from the local market and used on the crop. But in sum, growing cauliflower for commercial purpose is so far restricted to a limited people.

SOME COMMON PESTICIDES USED IN TOMATO, CAULIFLOWER AND CABBAGE

Some common pesticides used in tomato, cauliflower and cabbage are sulpher at the rate of 20kg/ha, borax foliar spray at the rate of 0.3% and sodium molybdate at the rate of 0.05% (RRTTS report, 2003).

POTATO CULTIVATION

According to some respondents, Bandhagada people in Phiringia block in Kandhamal district first introduced a few local varieties of potato during 1970s[2]. This gradually came to Katingia panchayat with the help of a school teacher who was appointed in Sudhipara village. He instructed the school children to cultivate this in the school garden. This gradually reached to the people of the villages such as Sudhipara and Laburi. Mallick families of Sudhipara village are the first to learn potato cultivation practice from the school teacher. They cultivated in a small portion of land with sandy soil (*balu bira*) and followed the procedures as instructed by the school teacher. This produced good result. Now 18-20 households in the villages are cultivating potato in the villages. Local variety of potato is small in size and sweet in taste. The maximum tuber size reaches 75-90 g. This variety contains more starch than the common variety. Like the beans the demand for local potato variety is also high.

The Kandhamal weekly market shows the uniqueness with the dumping of local vegetable produce such as turmeric, French beans, potato and others. Turmeric, French beans and potato are most common vegetable items sold in the weekly markets. Weekly market is also called as *haat* has a significant position in the Kandhamal economy. Since before independence, this place is a focal point to perform economic activities and other social and cultural activities. Non economic activities like negotiation of marriage, settlement of conflicts, exchange of information, recreational activities, sharing of ideas and thought about new agricultural practices are observed in *haats*. Vegetables, spices items, minor forest produce are some of the important produce of Kandhas come to the *haats*. In *haats* it is observed that apart from local people, many outsiders come to buy local potato varieties.

TURMERIC CULTIVATION

Mean Turmeric Production (in quintals) in the Villages

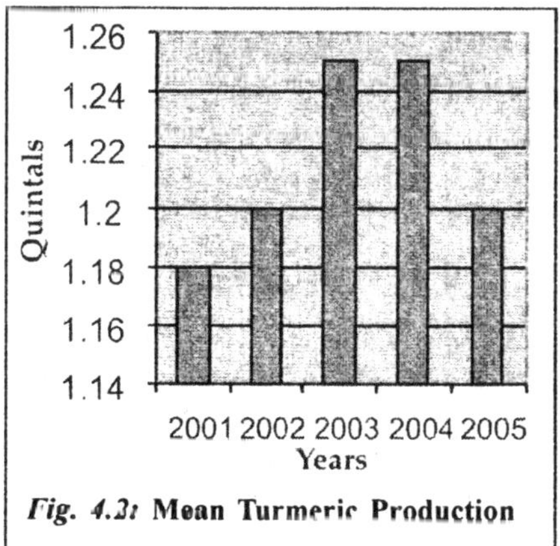

Fig. 4.3: Mean Turmeric Production

As mentioned in the previous chapter, turmeric cultivation is an age-old practice among the Kandhas. Turmeric cultivation is most favourably accepted by many people in the region. However, with the changing time, the crop is cultivated in a different form in recent period. Though majority of farmers use traditional seeds for the cultivation a few use new varieties of seeds supplied by the private agencies.

The above diagram shows that the average production of turmeric varies from 1.0-1.3 quintals during 2001 to 2005 in the villages. During 2003 and 2004 the average quantity of production in the village was 1.25 quintals, which is highest in five years. According to farmers these two years had normal rainfall and the number of households cultivated turmeric was more. The data collected from households shows that the difference in number of households cultivating turmeric is not much as compared to previous years. In 2002 the total number of households cultivated turmeric is 92 and in the year 2003 it was 96 and 98 respectively. The other important reasons for the fluctuating mean production are:

1. Variation in the use of fertiliser;
2. Quantity of seeds used; and
3. The variation in area under cultivation.

Sometimes the farmers take decision to replace some portion of their turmeric cultivating land by vegetable cultivation and *vice versa*. So the area of cultivation shows a little variation. The average land per household under turmeric cultivation varied from 0.25 acres to 0.30 acres between 2003 and 2004. It is also observed that some people use bio fertilser and some use chemical fertiliser in their turmeric field.

CHANGING TURMERIC CULTIVATION

The major changes observed in turmeric cultivation are the use of new technologies introduced by the involvement of public and private agencies. There is use of fertilizer among a few households. At present 10-12 households are using chemical fertiliser in the turmeric field. The common fertiliser urea is being used by the farmers for turmeric cultivation.

Raised bed planting is the new feature in turmeric cultivation. The households who use hybrid seeds prefer raised bed planting. But still mulching with traditional sal leaves is common. The private agencies like KASAM and SAMANWITA provide some incentives to the improved variety users; as a result, some people are inclined towards using so called improved varieties. But in general, the people prefer local varieties.

Rationale behind the Preference for Local Varieties of Turmeric

People are rationale in using local seed rather than hybrid seed. The reasons are as follows:

- People realized that the use of different types of soil for cultivation of local turmeric does not affect much to the yielding.
- In local varieties, the traditional method of cultivation is more suitable than the improved varieties. Raised bed planting, mulching and other methods are as such not required for the traditional varieties.
- In local varieties, the fear of pest and insect infection is less due to better resistance of the crops. The most important aspect is that the yielding from local varieties is better and the produce can be stored for longer time than imported varieties.

Despite some better qualities of traditional turmeric varieties, the emphasis is led to promote hybrid varieties. As said by the promoters, the hybrid varieties are promoted with the reasons that they generate good colour and aroma in comparison to the traditional varieties. However, farmers, by and large, do not agree with this.

Paddy Cultivation

Paddy cultivation is not new in the region. Since centuries, both the Kandhas and Panas have been cultivating paddy as main staple crop in the region. Paddy along with turmeric retained a special significance since time immemorial. Now, more than 90% farmers cultivate paddy in the region. Though a large number of households cultivate paddy, it is more or less confined to domestic consumption.

CHOICE, ACCEPTANCE AND REASONS FOR ACCEPTANCE OF CERTAIN VARIETIES OF PADDY

The trend of agricultural practice shows that paddy cultivation has retained the same value among the farmers. However, the local seeds have more or less been replaced by the hybrid seeds. It is worthwhile to discuss why there is high acceptance of HYV seeds in the region. The HYV seeds are not at par with local seeds as far as tastiness is concerned. But, due to climatic change followed by low or heavy rain fall, growing population, and change in consumption pattern, people have readily accepted the HYV seeds which are better adaptable to the environment. The growing research on paddy varieties by the regional and state research centres have enabled the farmers for successful adoption of HYV seeds. There are more than 100 high yielding varieties of paddy seed available in the state. However, only a few have been adapted in the Kandhamal agro-climatic zone. Short duration varieties and medium duration varieties of paddy are well accepted due to the dry climatic zone in the Kandhamal region. Short varieties with less than 99 days of the yielding capability help the farmer to yield early. Moreover, Jajati and Lalat varieties are resistant to pest and insect. These varieties are also suitable to local climate. The taste is also not bad for consumption. All these factors motivate the farmers to adopt the varieties.

At present, Lalat and Jajati are the important high yielding varieties, which have been successfully adopted in the region. Some other high yielding varieties such as Masuri, CR-1009, and CR-1014 have not been successful in the region.

It is not the climate alone that determines the adoption of certain varieties, but the availability, and easy technical procedure to follow the practice, often determines the acceptance by the general masses in the region. As observed in the region, the people do accept Jajati and Lalat varieties due to their easy availability in the region. Therefore, there has been wide access to these varieties.

It is also observed that consultation is important for adoption of new crops or varieties, which the farmers are not familiar with. In the region, some crops have become failure due to lack of proper consultation and training. As suggested by the agricultural scientists, CR-1009 and CR-1014 varieties are suitable in the Kandhamal agro-climatic zone, however, the farmers do not agree. The reason is that, in late nineties the above varieties were supplied to the farmers on an experimental basis. The farmers accepted the seed given free of cost. The fertilisers such as urea, potash, etc. were given free of cost. But, there was no proper training programs and orientation given to them to follow the systematic procedures for the practice. The farmers followed the same procedure as they followed in the traditional varieties. The lack of experience in the cultivation of new varieties did not help the farmers to produce good result out of it.

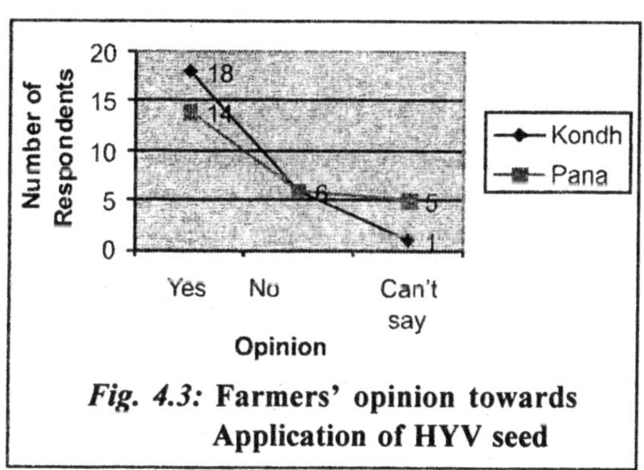

Fig. 4.3: **Farmers' opinion towards Application of HYV seed**

From an interaction with 50 farmers (25 from each category) to understand the opinion of farmers to know whether or not they like HYV seeds of paddy, could show that both Kandha and Pana farmers have positive opinion towards HYV paddy than local one. Out of 50 respondents, 32 farmers (18 Kandha farmers and 14 Pana farmers) opined that HYV seed is better not only for yielding but adapting to the low rainfall region (Fig. 4.3). The HYV seed of short variety is ripened within 80-90 days and production of this variety doesn't need enough water. However, only 12 farmers, six from each category, did not agree with the opinion that HYV seed is better. According to them local variety is better than HYV, as the former is more tasty and can be used for puffed rice, *lia*, and even *khir* (local food items). According to them there are also some short duration traditional varieties which give better yield. They pointed out that the HYV seed produces better quantity but not better quality like taste and aroma.

QUANTITY OF PADDY PRODUCTION IN THE VILLAGES

It is observed that despite the use of HYV varieties quantity of paddy produced is not enough as expected in the region. The reason is that per-household landholding in the region is less than 1 acre with average household size 5. As the agriculture is mainly rain fed, the paddy cultivation is done only once in a year. Again, there is less investment in paddy in comparision to vegetable cultivation. So the quantity of production in paddy is very less with mean production varies from 3.9 quintals to 3.1 quintals in the five years (2001-05) and standard error of mean .46883, .42509, .44077, .45130 and .41498 respectively. The figure as shown below does not show normal distribution of paddy production, but with huge variances in production. The above table shows that less than 4 quintals per household in the region are not enough to feed the population.

KANGA (PIGEON PEA) PRODUCTION

Kanga or *Kangul* (Pigeon Pea), a traditional pulse variety is now rarely cultivated. The restriction on podu cultivation has denied many people to continue kanga cultivation in the region. As the black soil of podu land and mountain land is suitable for growing up of *kanga* crops, the cultivation of the crop was more during pre-independence period. At present, the plain land does not provide adequate nutritional requirements for growing up of the crop. Some Kandha farmers opined that in present days kanga production does not give good return. It could be observed from the field that a few households had grown kanga in their kitchen garden, but the plants didn't not grow properly even during the time of flowering of plants. The pods size seems to be not properly grown and looking curly shaped. Some farmers pointed out that the soil is not so fertile unlike in the podu field. There is chance of pest attack to the pod, locally called as *jaurog*. As a result, there is drastic decline in the *kanga* or *kangul* production in the villages.

Table 4.4: Year-wise distribution of paddy production in the villages

(A statistical outline)

Year	*2001*	*2002*	*2003*	*2004*	*2005*
Mean	3.9000	3.7167	3.8264	3.8719	3.3636
Std. Error of Mean	.46883	.42509	.44077	.45130	.41498
Std. Deviation	5.13580	4.65658	4.84842	4.96426	4.56481
Variance	26.37647	21.68375	23.50713	24.64387	20.83750
Minimum	.00	.00	.00	.00	.00
Maximum	25.00	23.00	22.00	25.00	25.00

MEAN KANGUL PRODUCTION IN THE VILLAGES

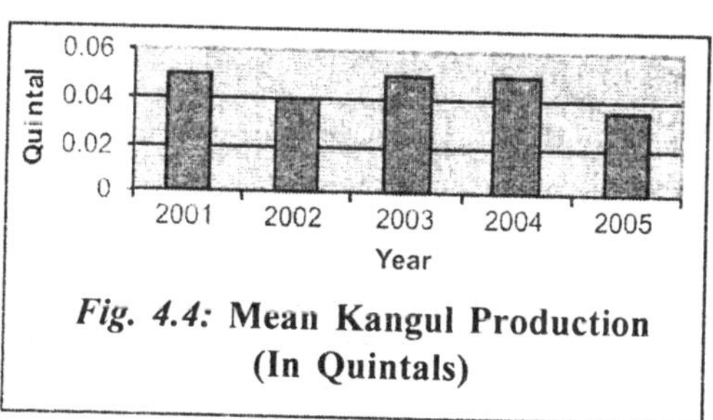

Fig. 4.4: **Mean Kangul Production (In Quintals)**

Figure 4.4 shows the mean distribution of kangul production for a period of five years. This shows that the average production of kangul varies between 0.02 to 0.06 quintals among the Kangul growers, which is very less in Kandhamal district. As already pointed out that this was one of the prime crops in the region and particularly in the villages, the decline of kangul production is definitely a concern in the region. In addition to the above reasons of decline in kangul production, the other reasons are that crop protection measures were not followed by the farmers and the agricultural agencies as well. Over the years, there is much emphasis on vegetable production but not for the conservation of traditional crops.

Table 4.5: Application of modern technology on crop in the villages

Crops	*Fertiliser*	*HYV seeds*	*Insecticide/ pesticide*
Paddy	Syamala, Gromor, Super potash, DAP	Lalat, Jajati, Konark, Kalinga-III, Surendra Khandagiri, Puja, Masuri	Nil
Vegetable			
Brinjal	NPK	BB-11, BB-1, BB-49	No use of chemical fertiliser Malathene
Potato	N.A.	E-4436, JF-27, JH-222 and local variety	
Beans (French beans)	Bio fertiliser/ compost (khat)	Mainly local	Nil
Turmeric	Mainly mulching with Sal leaves but a few use Shyamala, Urea, Ammonia	Mainly local Others are: Rashmi, Subarna, BSR, Rajendrasonya, Kasturi, etc.	Nil

IRRIGATION

It is found that 53.7% households are having possession of irrigated land in the region. But the important thing is that more than 95% of land is irrigated by spring water, which is seasonal. Only 5% of the total irrigated land is supported by canal irrigation. And, this 5% land is possessed by the Kandha households in Laburi village. The observation is that most of the land is supported with the spring water rather than canal or any other modern irrigation source.

Both Kandhas and Panas use local institutional mechanism to utilize and distribute spring water among all the land users. Local organizations in the form of village committee, Yuvak Sangh, and Self-Help Group (SHG) are actively participating in village development programs. Water management is an important aspect observed in the village. Management of water and irrigation not only focuses on the sharing of water but also focuses the effort towards the clean and chemical free(pesticide an insecticide free, fertilizer residue free water) water availability in the region. There is restriction, but limited, among the villagers not to use hazardous chemicals in the field in the region where the people use the water for domestic consumption. According to Pramila Pradhan, a woman farmer in Sudhipara village, "we have hardly seen any conflict due to water sharing for domestic as well as agricultural purpose in the entire region. Though there is water scarcity for some time, everything is managed well by the community."

During past few decades, government has been trying to construct canals, develop minor irrigation keeping in mind the growth of agriculture in the region. People have been constantly demanding for the development of irrigation facilities in the region as there is demand for water for growing vegetable crops through out the year. Following the demands from the people in the region, a small check dam and a watershed project materialized in the region. But this limited support has not completely helped the communities to overcome the water shortage during dry season.

From the use of matrix-scoring technique to understand the most urgent issues and the demands of the people in the villages it is noticed that the people rank irrigation as the basic issue in the village. Majority of the people demand for immediate solution of water problem to go for cultivation.

Table 4.6: Problem ranking through matrix scoring technique

Problems	*Sudhipara (group of 5 members)*	*Sudhipara (group of 6 women members)*	*Laburi (group of 4 members)*	*Laburi (group of 6 women members)*	*Total Rating*	*Rank*
Lack of irrigation	1	1	2	1	5	1
Supply Agricultural implements	3	2	2	2	11	3
Provision of credit	2	3	1	2	9	2
Fertilizer and other pesticides supply	4	6	5	7	22	5
Agricultural extension	5	4	3	4	16	4
Market problems	6	6	9	8	29	6
Others	7	5	4	6	22	5

It is noticed that lack of irrigation is the most important problem followed by provision of credit, supply of agricultural implements and agricultural extension. The discussion with the farmers of different groups, both men and women groups provides the logic why do people rank irrigation as the major problem. In fact, there are some traditional water sources such as spring and a small water reservoir in the region. The availability of the water is for seven to eight months (i.e., June-July to Jan-Feb). People use the water for irrigation purpose. There is no mechanism to reserve this water so far. Now, there is an urgent need to store this water for effective use in agriculture through out the year. The involvement of government officials is required to get this activity done. As the people are interested to cultivate different crops through out the year, water has become an essential component than any other inputs. In this regard, irrigation has been the major requirement. However, despite all this, there is not much effort made from the irrigation department and there is less government support extended to the region to tackle the issue.

FERTILIZER APPLICATION BY THE FARMERS

There is a significant change in the fertilizer application in the agricultural fields during last two decades. For centuries the Kandhas believed that propitiation of natural and supernatural beings can control the fertility of the soil. Therefore, there was practice of human sacrifice, locally called as Mariah in many parts during 18th century and before that. Moreover, people relied on the supernatural power for good weather, good harvest and so on. In due course of time, the traditional beliefs are superseded with their wide contact to the outsiders. However, Kandhamal is still at the bottommost end in use of chemical fertilizer. The average fertilizer consumption per hectare in the district is around 3 kg, which is much below the national average of 86.34kgs per hectare. In the villages the average per hectare fertiliser consumption is 15 kgs, which is higher than the average of the district, but less than the State's average of 39 kgs per hectare (Directorate of Agriculture and food Production, Orissa, 2001, Kandhamal District Agricultural Statistics Report, 2000-01, *Agricultural Statistics at a Glance*, 2003).

It is noticed that the Kandhas are reluctant to use chemical fertilizer in paddy field because they believe that the use of it affects the nature of purity and quality of soil. Particularly, the Kandhas are more conscious in retaining the soil fertility of the region. They are forerunners in using bio-fertiliser such as compost, cow dung, etc. Similarly, the Panas too come forward to use bio-fertiliser to their limited agricultural practice. But the basic difference between these two communities is that the Panas are not more conscious about the longevity of soil fertility and their sustainability. They use fertilizer which is easily and cheaply available irrespective of kind and composition.

In general, the entire population of the region have positive attitude towards bio-fertiliser. A survey of 50 farmers (both Kandhas and Panas) to understand the farmers response towards application of bio-fertilizer and chemical fertilizer shows that majority of farmers (56%) are in favour of applying bio-fertiliser than chemical fertilizer in the land. Again the percentage of Kandha respondents favouring bio-fertiliser is more than the Panas. 60% of Kandha farmers and 50% of Pana farmers are in favour of using bio-fertiliser. Again, only 30% of Kandha farmers and 50% of Pana farmers prefer the application of chemical fertilizer. So the percentage of the total farmers preferring chemical fertilizer application is 34%. And, there are 10% of total farmers not able to answer as which one is favoured for the use. In further analyzing the aspect about the farmers' preference to bio-fertiliser, it is understood that this fertiliser is easily available, cost-effective, can retain the longevity of the soil quality and water retention capacity of the soil. In addition, it is easy to apply also. For majority farmers, the use of bio-fertiliser does not affect taste of the produce than the application of chemical fertilizer. The produce can be stored for longer time if there is no use of chemical fertilizer. The use of chemical fertilizer, in other ways, is very expensive, difficult to access, reduce soil longevity, requires much water for the crops to cope up with the chemical fertilizer, and so on.

Again, it is noticed that the small landholders having sound income to invest prefer chemical fertilizer for better productivity. They emphasize on increase in productivity per acre. A very few big and medium landholders emphasize on the application of chemical fertilizer. The small landholders and vegetable growers prefer the use of chemical fertilizer. They pointed out that chemical fertiliser is helpful to give instant result. It also helps the crops to produce more out of a small quantity of land. However, they say that bio-fertiliser is important for cereal production. Bio-fertiliser is prepared by accumulating cow dung, extracts of household waste materials, straw, leaves and others. It needs time to prepare. If during that time chemical fertiliser is available to them they like to go for that. According to them, though the use of chemical fertilizer is expensive, but it can be used on a limited land and restricted to certain crops such as vegetable. According to them, chemical fertilizer is must for vegetable production in such case when there is no availability of enough bio-fertiliser. Some respondents say that they could prefer bio-fertilser if it is readily and promptly available for them.

RESPONDENTS' VIEWS REGARDING USE OF FERTILISER

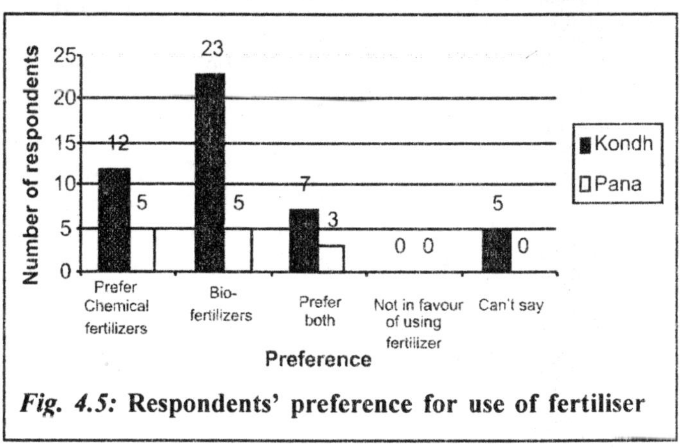

Fig. 4.5: **Respondents' preference for use of fertiliser**

In order to understand the views of farmers regarding the use of fertiliser, it shows that out of 40 Kandha farmers interviewed, as shown in the adjacent diagram, 23 respondents prefer bio-fertiliser application to chemical fertilizer, 12 prefer chemical fertiliser application, seven Kandha farmers have equal preference to both, and five respondents are unable to answer about their preference. Among 10 Pana farmers interviewed, the number of respondents preferring chemical and bio-fertiliser is same. For many, bio-fertiliser is better in long term, but for quick result the use of chemical fertiliser is preferred. Again some farmers, those are in favour of using fertilizer, are of the opinion that the chemical fertilizer is necessary for vegetable and newly introduced crop varieties. However, in traditional crop varieties they do not prefer the use of chemical fertiliser.

Rise of cooperative agencies and emergence of extension centres in the region are now the important factors for the promotion of fertilizer application in the region. It is found that some of the farmers are using fertilizers such as Urea, NPK, Super potash, and Shyamala in vegetable field and rarely in paddy field.

USE OF BIO-FERTILISERS IN AGRICULTURE

The use of organic fertilizer in the field is not new in the region. Since time immemorial, the farmers have been cultivating their land and practicing agriculture without the application of any chemical fertilizer in the region. Except certain villages such as Sudhipara and Laburi villages there is very low chemical fertilizer application. Over the years the production practices in agriculture continues without application of any fertiliser. Still many people in the district are unaware of chemical fertiliser application. There are many regions in the district where still chemical fertiliser and pesticides are not known to farmers. There is effort made by the government to declare the district as organic zone. This in due course of time has attracted wide attention from both government and non-government agencies to promote organic agricultural practices further. International agencies like, SKUL of Netherlands and ISOFAR have taken some initiatives to promote organic farming in the district.

Some self-help groups have come forward to promote growing of turmeric without application of any chemicals in the region. Agencies like KASAM and SAMANWITA promote groups from among males and females to develop interest for organic agricultural practices.

LIVESTOCK REARING

Livestock rearing has a special importance in the district. Being a backward district in the State of Orissa and having more than 80 per cent population dependence on agriculture, the importance of cattle and other livestock is more in the district. As per livestock census (2003) the distribution of livestock population in the district is given below.

Table 4.7: Livestock distribution in the district (in Numbers)

Cattle	338127
Buffalo	68277
Goatery	212383
Poultry	428384

These varieties of livestock are mainly indigenous or local varieties. The majority cattle are non descripiptive type with few ghumsar and upgraded progeny of jersey, Holstein brown swiss and red shindhi etc. The buffaloes are also nondescriptive with upgraded progeny of murriah buffaloes. Goats are mostly boatel, Graded Kalling (Ganjam) and Black Bengal breeds. Feeding, housing and breeding are traditional and natural in all livestock cases (RRTTS, G.Udayagiri, 2003).

In Sudhipara and Laburi villages, the people have generated interest to own cows, bullocks and buffaloes for the ploughing and for the collection of dung for the fertilizer purpose during the recent years. Cattle rearing have increased in recent years with growing interest on vegetable production in the region.

Table 4.8 shows that in the villages there are only 7 households having ownerships of goats, among which four households belong to Panas. However, the ownership of other cattle shows that the Kandhas own more cattle than the Panas in the region. The agricultural communities have to rear the cattle for timely cultivation. Cow milk is not preferred in the villages. The Kondhs engage both cows and bullocks for ploughing. Buffalo is the most preferred one. but it is comparatively expensive. Purchasing of one buffalo costs around Rs. 10,000 to Rs. 12,000. For the same cost a pair of bullocks can be purchased. Therefore, the ownership of buffaloes is less than bullocks in the villages.

Table 4.8: Livestock population and distribution among Kandhas and Panas in the villages

Ownership	*Social Groups*		
Bullock	Kandhas	Panas	Total
Less than one pair	2	1	3
One pair	30	4	34
More than one pair	14	0	14
Nil	43	24	67
Cow	Kandhas	Panas	Total
Less than one pair	9	6	15
One pair	13	2	15
More than one pair	6	2	8
Nil	61	20	81
Buffalo	Kandhas	Panas	Total
Less than one pair	1	29	30
One pair	1	0	1
More than one pair	6	1	7
Nil	81	29	110
Goat	Kandhas	Panas	Total
Less than one pair	1	0	1
One pair	0	1	1
More than one pair	2	3	5
Nil	87	26	113

The observation is that the Kandha people do not prefer to rear goats as they are a threat to crops. According to them, goats, usually, destroy the fencing and enter into the agricultural field. But Panas do like to rear goats because they are mainly landless people and goat rearing is a good business. In certain cases, the goats which are often reared by the Pana families enter into the agricultural field if they are not properly checked. Keeping the above into consideration, the farmers have followed strict and rigid rule to check the farmers in the region. Irrespective of Pana and Kandha everyone has to obey the rule. Fine has been imposed on some families for the above action. Any violation of the rule by a member is punished with fifty rupees fine.

OTHER AGRICULTURAL TOOLS AND IMPLEMENTS

As far as the use of other agricultural tools and implements is concerned there is a sign of change in the region in comparison to the pre-independence period. Dependence completely on traditional agricultural tools and implements is a story of the past. With the advancement of time, the use of modern tools and implements has become essential for the farmers in the region. With the change in crop production, the change in use of agricultural implements has become essential for the farmers. In present agricultural system, the able-farmers are showing interest in buying iron ploughs, sprayers/sprinklers, winnowers, threshers and other mechanised tools and implements.

From the primary fieldwork it is substantiated that nine households have iron-plow, eight households have sprinklers or spraying machines, and one household owns a tractor. There are two paddy winnowers and two threshers owned by the Sudhipara village supplied through agricultural agencies. It is observed that all the modern agricultural implements are owned by the Kandha households except two paddy winnowers and two threshers that were supplied by the extension centre of Krishi Vigyan Kendra (KVK) of G. Udayagiri.

USE OF SPRAYER IN AGRICULTURE

The use of sprayer is new in the region. In the year 2002, first, one Kandha household acquired the sprayer with the support of subsidy supplied by the agricultural office. In a period of 3-4 years the importance has been realized by many households; mainly, agricultural households are showing interest on acquiring the sprayer. Though the awareness was there before

2002, there was not such requirement. The use of insecticide and pesticide was very less in the previous decades. However, there is growing realization after the frequent affect of pests and insects to certain vegetables like potato, beans (French beans), and tomato crops. The rottenness of potato tuber, curliness of leaves of tomatoes and potatoes, etc have made it compulsory among the farmers to use pesticide and insecticide, though people never used them before. The precise reason for the growing pest and insect attack on crops is unknown to farmers.

PESTICIDE AND INSECTICIDE

According to the people, during past few decades there is growing concern about crops affected by diseases in the region. Particularly, after taking up of growing vegetable crops, this problem is faced by the farmers in the region. Therefore, the use of sprayer has become essential in the region.

Table 4.9: Some common plant diseases in the region

Name of the disease	*Characteristics*	*Plants affected*	*Seasons when the disease prevalent*	*Local coping mechanism*
Kadidipka	Curly leaves, whiteness of leafs	French beans, Lablab beans and other legume varieties	No specific season	No specific mechanism
Shaiba	Rottenness of tubers	Potato, ginger like tubers		No specific mechanism
Sanikakte (Jaurog)	******			No specific mechanism

Despite some common plant diseases in the region, the people have no specific local mechanism to control these pests and insects. Their knowledge to control pests and insects is very less. In view of this, the use of pesticides and insecticides available in the local market has become essential.

But, some people respond negatively about the use of pesticides and insecticides on their crops. Mainly the elderly people in the region oppose the use of pesticides and insecticides. According to them the vegetables lose the taste and quality after use of pesticides and insecticides. Some people are aware of air pollution too. They view that the use of chemical pesticides or insecticides pollute the air and this pollution affects their health. But this awareness might have developed due to increase in education, contact with urban people and also interaction with agricultural research centres.

Some elderly people are not happy with the present cropping pattern and use of modern technologies. Their claim is that the increase in plant diseases is due to use of modern tools and technologies. 'Rama Krishna Pradhan', a seventy-year old farmer said:

> "there was no concern of crops affected by pests and insects as we used to use only local seeds, local tools, and local crops. Because of the use of modern technologies like HYV seeds, fertilisers, and chemical pesticides and insecticides the food is tasteless and problem to our health is more. The use of modern technologies has removed our traditional agricultural practice in certain sense."

Though the region has shown some receptivity to adoption of new agricultural technology, still there is varied response among the farmers with regard to adoption of modern inputs. For the farmers having vegetable cultivation, the use of fertilizers, pesticides, and insecticides is an essential thing; while for the traditional paddy grower, the use of the above is not essential. Small landholders and vegetable growers need the use of modern technology more urgently to increase the productivity. The response in the use of modern technologies varies on the basis of differences in age groups. The middle-aged people (above 35-50) emphasize the use of bio as well as chemical fertilizer. However, people with more than 80 years of age do not prefer the use of chemical fertiliser in the agricultural field. Their inclination to traditional practice and technologies is more.

PADDY WINNOWERS AND THRESHERS

Paddy winnowers and threshers are other new agricultural implements used by the farmers in the region. However, these instruments have not been so popular among the farmers. Because, the threshing and winnowing job is basically handled by the women in the region. They do it so happily. It is not a difficult task but rather an enjoyable activity for the women in the community. Threshing activity is mainly done without engaging cattle unlike in some plain areas in the district where the Oriyas engage cattle mainly for threshing of paddy and similar other crops. Both men and women engage in threshing crops. They do it manually. Use of a wooden thresher is seen being used by women. Sometimes, this activity is done by hand or foot. So introduction of thresher and winnower is not so important for the agricultural supports.

USE OF MECHANICAL INPUTS LIKE DIESEL PUMPS AND TRACTORS

Use of diesel pump for irrigating crop field is noticed among the farmers in the district now days. Wide popularity of vegetable cultivation in the region has attracted some farmers to use of diesel pump sets. For the progressive farmers, diesel pump is a status symbol. In the villages there are four diesel pumps owned by the people. One diesel pump set is owned by a Pana household and the other three are by Kandha households. Sometimes, these diesel pumps are rented to other families who need them for irrigation.

Though tractor is an important component of agricultural modernization, the use of tractors in agricultural fields is rare in the villages. There is a tractor owned by one Kandha household in Sudhipara village, which is mainly used for renting. At the time of field study, the charge to hire a tractor was Rs.500 per day. The same amount is also charged to plough one hectare of land. During road work, house construction, and other construction purposes, this tractor is also hired from him. Though there is no instance of using the tractor for agricultural purpose in the villages, some tribal land lords from other villages hire it for agriculture purpose.

THE FACTORS AFFECTING TRADITIONAL CROPS PRODUCTION

Despite the continuation of mechanization in some agricultural activities, the use of traditional technology continues in the region. Still there is importance of *kodi, koruwa, gadi, rampa, pikash, langal* and other traditional agricultural implements for agricultural activities.

But, in due course of time many traditional crops such as *jower*, maize, *kuiri*, etc have lost the place due to various factors. Firstly, enforcement of ban on the traditional podu cultivation has been the major cause of taking up new agricultural practices. It has also caused loss of traditional agricultural occupation by many farmers in the villages. Facing these issues, most of the families searched for new job opportunities for survival. In this situation, settled cultivation emerged as a major support for the people in the region. Following this, vegetable cultivation entered as an alternative in the region. The enforcement of restrictions on podu cultivation, farming innovations of a few farmers, interaction of these farmers with other farmers in the district, etc resulted the community to opt settled cultivation and latter to take up vegetable cultivation more aggressively.

Secondly, the adoption of cash crops gradually led the farmers into using new agricultural inputs. Use of new seeds and chemical fertilizer became essential after 1980s with growing adoption of vegetable cultivation. The people started focusing on the new crops. Their focus on traditional crops gradually reduced. Most of the time both males and females engaged themselves is the new agricultural practices.

Thirdly, the market was not favourable for the continuation of traditional practice. The traditional agricultural produce did not fetch reasonable price to the farmers from the local market. The time and labour consumed for cultivating this produce is more and needs sufficient labour cost. But the return is less. Even the public and private organizations were not so supportive for continuation of the traditional practice During post independence period the research and extension focus was only on cereals like paddy and wheat and cash crops like vegetables. There is less effort on protection and development of traditional cereal crops like Jower, and millets like *kuiri, arca, klingiraka, etc*. This directly or indirectly affected the continuation of the practice.

THE ROLE OF OTHER FACTORS

All the major changes observed are achieved within two-three decades. Increase in literacy level, laying of pucca roads into the villages under "Pradhan Mantri Gram Sadak Yojna" (PMGSY), transport facilities, and other factors, have brought people into the fore of urbanisation. These factors play crucial role tin the adoption of mechanized inputs. It is observed that during recent years the people are more or less able to save some amount for the future. This was not very common in the past among the tribal societies. However, with the influence of outside world, there is change in behavior. More particularly, the women folk have induced their husbands, at least, for some saving for future investment in agriculture, education and other purposes. During the discussion with an 'Anganwadi Worker', it is found that the alcoholic consumption habit among the Kandhas has been checked, which reduced unnecessary expense. The Kui Samaj, a registered society, has checked the alcoholic consumption in the Kandhamal district. This has rescued many households off their debt burden. Similarly, growing 'Vaishnavism' in the district is also one of the reasons for giving up liquor and other alcoholic drinks. More than 25% households in Sudhipara village have embraced this spiritualism with the influence of Shri Ram Das, a Hindu spiritualist in Ghumsar region. There is also influence of '*Vanabasi Sangh*' who promotes spiritualism among Kandhas in the region. This *Sangh* has created awareness to stop consuming liquor mainly among Kandhas in the villages.

In sum, it is observed that during recent years some of the traditional agricultural practices are being continued and several are being replaced with new practices. There is change in land ownership, crop production, fertiliser and HYV seed application, etc in agriculture. Some traditional crops such as sim, turmeric, kangu exist with variation in quantity of production. But the introduction of new cash crops such as vegetables and other spice items are increasing over time. Application of fertilizer is relatively new phenomenon. There is still continuation of traditional tools and implements in agriculture. The Kandhas use traditional tools like *pikash, langal, kanti, korada,* etc in the agricultural practice. Meanwhile, there is use of diesel pumps, winnowers, threshers, spraying machine like mechanized inputs. But engagement of human labour is still of primary importance. Women continue to play significant role in agricultural operation. This shows that the agriculture is in transition. There is neither complete modernity nor complete tradition in this region, but there is change over time.

It is also found that the farmers, by and large, are progressive and have modern outlook. They are aspiring to access new innovations. They select the crops which can better adapt to the local climate and best suited to the local social and natural environments. Market and infrastructure are also some important preconditions in addition to environmental factors. Selection of crop variety is based on consumption pattern, local demand, market support and quality of product as well. Over the decades some crops have been successfully adopted and some have been failure in the region.

REFERENCES

1. District Statistical Report, 2005.
2. It is said that the Bandhagada people in Phiringia Block Brought this Dwarf Variety of Potato in the District. But some say this Variety Might have come from Bhanjanagar in Ganjam district. Bhanjanagar and Berhampur are the Important Trading Centres in the Ghumsar Region, which includes G.Udayagiri also.

❑ ❑ ❑ ❑ ❑

CHAPTER - V

EFFECTIVENESS OF INVOLVEMENT OF PUBLIC AND PRIVATE AGENCIES IN AGRICULTURE

The role of public and private agencies has been significant in delivering various services to the farmers both during and post green revolution period in India. Transfer of knowledge, technology and innovations to the farmers has been crucial both in rural and urban areas. Over the years, the researchers, extension officers and voluntary workers have been trying to develop agriculture in backward areas in various ways such as providing employment opportunities, improving horticultural practices, creating self-help groups (SHGs) and providing necessary inputs like technology, finances, and other necessary supports. However, the convincing result is yet to be materialized. It is realized that the farmers in the remote and backward areas are lagging behind in accessing the services delivered by both public and private agencies.

The chapter sketches out the role and effectiveness of public and private agencies. It highlights issues of farmers in the perspective of involvement of public and private involvement, technology dissemination and agricultural production. Moreover, the chapter deals with the issues associated with the technology transfer into the Kandha dominated regions where agriculture has become a primary occupation of more than seventy-five percent households. It focuses the effectiveness of the services rendered by the public and private agencies in agriculture in the tribal villages. Effectiveness refers to the ability to meet goals, objectives or needs (Carney, 1998). Here the goals, needs and objectives of the public and private agencies are to support the farmers towards better agricultural production in a sustainable manner. And this is possible by extension of effective services by supplying necessary inputs to the farmers. The chapter is divided into several parts in order to understand the effectiveness of involvement of public and private agencies in agricultural development programs.

I

First, the chapter reflects on organizational structure and institutional arrangement for agricultural development in the region. The institutional arrangement in the region includes an amalgamation of both the formal and non-formal organizations associated with activities for agricultural development.

FORMAL ORGANIZATIONS

Formal organizations are based at the District level, the Block level, the Panchayat level and the village level. The structure is a three-tier one. Firstly, the public agencies, secondly, the private agencies and thirdly, the local and community based organizations. All these institutions are interlinked and their contribution varies as far as the provision of services is concerned.

Public Agencies

The public agencies include agencies, which are run by both the state and central governments. It includes the formal administrative structure such as District Rural Development Agency (DRDA), Integrated Tribal Development Agency (ITDA), the Block office, the Tehsil office, etc; Research Based Organizations like Krishi Vigyan Kendra (KVK), Regional Research and Technology Transfer Station (RRTTS), Horticulture offices, etc, which are also the implementing agencies. Similarly funding agencies like NABARD, DFID and UNDP at the macro level, and at micro level the organizations like LAMPS, SBI, UCO Bank, etc for providing credit support are the part of agricultural development programs. These agencies have multifarious role in the agricultural development both directly and indirectly.

Technology Transfer Centres

There are research institutions viz., RRTTS, KVK, and Office of Additional District Agricultural Officer (ADAO) located in the region exclusively for the extension of technology in agriculture. Krishi Vigyan Kendra (KVK) and RRTTS are working under the Orissa University of Agricultural technology (OUAT) and mainly financed by central government. But, ADAO office is under the directorate of agriculture of the state government. The technical staff associated with these institutions are basically postgraduates in the field of agricultural science. The present situation is that there is only one specialized extension officer who is in-charge of both the institutes such as KVK and RRTTS. But, in ADAO office there is no specialized extension officer for rendering the service. The provision is that the ADAO office members can work closely with the Block officials like village agricultural workers (VAWs) and agricultural extension officer (AEO) for effective transfer of the knowledge and technology and supply of the inputs.

Soil Conservation Department

This department was set-up at G. Udayagiri in 1970. This institute is directly associated with the communities for improving the natural resources and soil conservation in the region. The main objective of the institution is to supply necessary inputs to farmers to protect the soil from erosion; regeneration of fertility of the soil, erosion control, check dam construction and run-off management, etc. Since its inception, the following activities have been taken up by the department:

1. Plantation or Afforestation;
2. Gully control;
3. Construction of check dams;
4. Podu prevention scheme;
5. Check dam construction and runoff management; and
6. Dug well construction.

Application of modern scientific knowledge is the base for developing following activities. Use of semi-mechanical method (SMM) is predominant in some of the activities such as gully control, construction of check dam, etc.

Plantation and Afforestation

The plantation and afforestation program is a supporting mechanism for development of forest in particular and environment in general. Linking agriculture and plantation, and afforestation program is important to provide immediate and alternative support to the landless labourers through wage earning and in long term, a sustainable environment for the agricultural development in the region

In 1975, as a part of the Draught Prone Area Programme (DPAP) activities, the coffee plantation program was introduced in Lingagada and Jakamaha villages of the then Lingagada Panchayat. According to the people this had created an income opportunity for at least one year through wage earning. Though people from many villages took part in the program in the villages, their involvement was mainly confined to wage earning. However, active participation of the tribals in the form of decision making, management, and conservation was very less. After the plantation there was little effort made to take care of the plants. As such, there was no effort made to create awareness among the people about the benefits of the program. Moreover, the podu cultivators who lost the land were not provided with any incentive rather they were considered guilty when they further attempted to continue their practice of podu cultivation.

The plight of the podu cultivation can be seen in the words of Sahadev Pradhan (age, 48 years) of Bijigam hamlet in Sudhipara village, "I had a patch of land on the Priki Saru hill. I used to cultivate the traditional varieties of cereals and millets, which was sufficient to meet the end. Ever since the forest officials and group members restricted me from cultivating podu land, I am unable to manage my family. I am now a poor landless laborer in the village. There is no alternative income opportunity provided to me. I am not against the plantation and afforestation program. But what about my livelihood?"

So the major problem of the program was not with the activities undertaken, but the required activities to provide alternative support for the livelihood were not undertaken.

Horticulture

Horticulture operations in the district go back to 1970. But in 1980s, the program was started in Katingia panchayat. Following up of which, commercial crops such as coffee and spices like cardamom, black pepper and cinnamon plants were cultivated in some parts. However, this activity could not generate peoples' interest as they were less aware of the programs. Moreover, the program was new to them. Devoting time in the coffee plantation than devoting in turmeric and ginger fields was thought useless by them. In addition, incentives were not supplied to the people to promote the program. Gradually, the people lost interest as they were not provided with any government support after the plantation work was finished. The community participation became very less due to lack of incentives to the members.

During a discussion with the people, it was realized that the tribals are not against plantation or horticulture but shifting suddenly from the traditional agriculture is a difficult task among the communities. Another important issue that could be derived is the level of awareness among the people regarding new plantation activity. Many said that there was no activity undertaken by the government regarding awareness creation about the usefulness of newly introduced plants. The people were less aware about the value of the new plants such as cinnamon, black pepper as these are completely new to them. Apart from that, it is also observed that people have their own interpretation about crops. When they compare the newly introduced cash crops with the traditional cash crops, the first thing that strikes to their mind is the irrigation availablity and local market support. However, for horticultural development water is as such not a big issue but there is very less scope for market in the region to take up this cultivation. The people need instant market support to the products they grow. They select such crop which can be sold and used conveniently without difficulties. For example turmeric, ginger, and chilly are preferred by the farmers because they can be used in different forms. Both raw and ripen products are used by the people for their own consumption and also for market purpose. Marketing support to these products is regular. Production strategy is known to almost all the farmers in the region. The local climate is suitable for most of the spices items as mentioned by the horticulture specialist. However, the social and marketing environment plays crucial role to determine the production preference.

The people consider various aspects before and after the cultivation. From the production to marketing every stage is important for them. The first important thing among the communities, in present context, is the institutional support like market. They also need support from the public and private agencies in the form of inputs supply, technical support, and financial support for agricultural development. The value of new horticultural products and cash crops though very high in national and international market, but this scope has not been chanelized through efficient market support in the region. The lack of market support did not encourage farmers to continue horticultural growing in the region. As a result both Kandhas and Panas freely allowed their cattle in the cropping areas. These cattle destroyed the fences, entered into the gardens and spoiled the plants. No organized meeting was conducted nor was any group formed in the region to request the people for not leaving their cattle into the areas.

The adoption of new cash crops should be suitable from geographical and climatic points of view. But the less awareness about the value of crops, rudimentary market support, stray cattle grazing problem, inadequate public support, etc are responsible for poor horticultural growth in the region.

Now, this situation has more or less changed. The people themselves are aware of the need of the modern cultivation. It is also observed that after 1960s, several institutions like regional agricultural research centres, soil conservation department, irrigation department, and several NGOs have been set up in G. Udayagiri block. Several programs and schemes have been implemented during last four decades. This institutional framework has also lead to change the agricultural scenario in the villages.

Gully Control

Soil conservation department took the initiative for gully control in 1976-77 in the district. But, the same measure was taken in the Katingia Panchayat in 1980-81. So far, there are only two sites at Katingia and Tiangia villages under gully control program.

In this program, the local knowledge also plays crucial role than the scientific knowledge. The community uses its local available resources for conservation of soil. One such example of local knowledge in soil conservation is explained below:

> 'In order to avoid the flow of fertile soil from the cultivable land, people use sal and bamboo poles in the form of fence in the risk prone area where the current of water flow is more. The poles are used between two or more soil ridges of upper and lower ones. Sometimes these poles are used in order to divert the water flow during rainy season. The poles are arranged in such a way that the current of the water flow is reduced and the fertile soil is restored without erosion'.

Dug Well Construction

Dug-well construction programme was initiated in 1970s. In the villages, there are three dug-wells constructed by the

Integrated Tribal Development Agency (ITDA), Baliguda in Kandhamal district. After taking up vegetable cultivation programs in 1970s the demand for dug-well construction has increased in the region. In the same years ITDA took up the initiative to construct dug-wells in the tribal region. But, so far, the achievement is very less. In 1990s the government policy shifted from individual focus to group focus. Now, there is drastic reduction in amount of loan and the number of beneficiaries in the region. After 2001 there is no instance of dug-well construction in the villages. There is increase in demand from the people for dug-well in the villages after taking up of vegetable cultivation. Despite that, dug-well programme has not been successful here. ITDA emphasizes infrastructure development like construction of road, irrigation, assisting school activities, etc. But as far as Sudhipara villages are concerned there is no such substantial activity undertaken by ITDA.

Construction of Check Dam

Construction of check dam was started in 1987-88. There were two check dams constructed at Tiangia of Katingia Panchayat in 1992-93 and 1994-95 respectively. From these two check dams, 125 hectares of land have already been irrigated. The new irrigation technology could draw the interest of the communities which was so effective for the vegetable cultivation as well as paddy cultivation. But the construction of two check dams is supportive to a few families in the Laburi village of Tiangia region and only 20% of families are able to access it. Where as, the rest 80% of the total families depend on rain fed agriculture.

Spices Board

The involvement of both the private and public agencies in spices promotion is significant in the region. In 1980s, horticulture department started supplying the new HYV/ improved variety of turmeric and ginger in the region. Following this, there was a decision taken by District Rural Development Agency (DRDA) to set-up a separate board for spices production in collaboration with the horticulture department. Similarly, KASAM, a private agency based at Phulbani was also associated in the board to take up the spices production activity in the district. The spices board was set-up in 1990s in G. Udayagiri block. This is one of the six spices boards set-up in Kandhamal district under the authority of DRDA. The other spices boards are Madingisur in Raikia block, Nuagaon and Chanchedi in Nuangaon block, Phiringia in Phiringia block, and Suganketa in Daringibadi block. The Chairman of the board(s) is the district collector. The board works closely with the regional agricultural research center and Regional Research Technology Transfer Station (RRTTS). There are secretaries appointed by the board chairman for smooth function of the spices board and creating interest among the farmers. The board chairman appoints the secretaries of the boards. The secretaries are mainly from the tribal community who are having organizational capabilities and communication skills to persuade people to get into the activity.

Activity of G.Udayagiri Spice Board

In Katingia Panchayat the total amount of improved varieties of spices supplied to the SC and ST households is 5.5 quintals till 2000. This includes 2 quintals ginger and 3.5 quintals turmeric. The total beneficiaries who received ginger are 355. Out of these beneficiaries, the Pana beneficiaries are 120 and Kandha beneficiaries are 235 in number. Similarly, total beneficiaries who received turmeric varieties include 540. Beneficiaries include 220 Pana households and 320 Kandha households. In the villages such as Sudhipara and Laburi the total seed supply includes 4 quintals of which 1.2 quintals is ginger and 2.8 quintals is turmeric. The total Kandha and Pana households receiving ginger and turmeric include 108 and 116 households respectively. Seventy-eight Kandha households have received both turmeric and ginger seeds. Thirty Pana households have received ginger seeds and 38 Pana households have received turmeric seeds (Table 5.1).

Table 5.1: The total seeds supplied in the villages from 2001-04

Region	*Spices*	*Quantity supplied in quintal*	*Pana (No. of households)*	*Kandha (No. of households)*
Katingia	Ginger	2	120	235
	Turmeric	3.5	220	320
Sudhipara and Laburi	Ginger	1.2	30	78
	Turmeric	2.8	38	78

Source: *Data from Katingia society collected from Loknath Pradhan,* the Ex-Secretary of Spices Board.

The Aim of Spices Board

The purpose of setting-up the board is to supply improved variety of seeds to the farmers for better production, supplying market opportunities and developing the technical skills through field visits, training, and consultation.

The main purpose of the above is enhancing livelihood condition of the marginal farmers in the region. Initially, each spices board was provided with Rs. 5,000-6,000 from the NABARD fund. The secretaries of the boards tried to persuade the people for their membership in the board. The farmers, who were interested, paid ten rupees as membership fee. After getting membership, the farmers received improved seeds. They were given free consultations at the training centres in G. Udayagiri. Initially five to six households in Suhdipara village took membership and introduced improved varieties of turmeric seed for cultivation. After harvesting, they returned the same quantity of seed to the board that they had received. There is provision in the board that if any farmer shows interest to sell the produce to the society or the board, he can sell at the price fixed by the government. The board members and secretaries under the leadership of the chairman decide the price of the product. Every year or in alternative year, the price of the item is revised based on the market price of the item. The members who are appointed as secretaries and members draw the attention of many farmers in the region to accept the improved varieties. The staff members convince the people by saying, "the new seeds are having qualities like better production, good color, more fragrance and more taste than the traditional varieties. The market price will be better and the duration of each cropping will be comparatively less."

According to Loknath Pradhan, the then secretary of the spices board at Katingia, the three main varieties that were supplied to the farmers included *Alipi, Surama* and *Lagdong*. Other than the improved varieties, the farmers were also supplied with fertilizers such as Shyamala, DAP, Urea, Super Potash and Ammonia at cheaper rate for applying in turmeric field. The farmers received the fertilizer but many of them did not apply in the field initially fearing the loss of original quality of the soil. However, the demand for the fertilser has gradually increased with the increase in importance of vegetable cultivation in the region. But in contrast the supply of fertiliser got reduced. In the last two years there was no supply of the above inputs to the farmer in the region.

Involvement of Private Agencies in Agriculture

The private agencies, on the other hand, have taken a crucial step to promote new technology in agriculture in the region. Private agencies like NGOs and voluntary agencies, and international agencies have taken part actively in agriculture development in the region. In the villages the voluntary agencies associated are SAMANWITA, KASAM, and PRADATA and the international agencies are UNDP, World Bank, Food and Agricultural Organization (FAO), etc. These agencies are working together with the government agencies. While the former funds the projects the latter provides the technical experts and manpower support with the objective to promote agricultural production in the region. Similarly, the voluntary agencies work with the public agencies such as block, Panchayat, Krishi Vigyan Kendra (KVK), and RRTTS for the collection of information, training, selection of areas and the samples for the project implementation. These NGOS and voluntary agencies also work with the Community Based Organizations (CBOs) to implement the project at the grass root level. Some of the CBOs are developed with the help of NGOs and voluntary agencies.

KASAM

'Kandhamal Apex Spices Association for Marketing' is an agency working for the promotion of agriculture and marketing in the region. It especially deals with the spices promotion among the Kandhas in the region. This agency was set up in 1983. Soon after the establishment of the agency, the promotional campaign of two major cash crops such as turmeric and ginger was taken up. The organization selected some of the influential youth to promote the campaign and convince the farmers in order to take up new varieties of ginger and turmeric that were imported from other states.

SAMANWITA

SAMANWITA is another agency closely working with public institutions such as District Rural Development Agency (DRDA) and funding agencies like United Nation development Programme (UNDP) for the livelihood and agricultural development in the region. Since setting up of this organization there have been many activities taken up by this organization. Developing self-help groups is the major activity of the agency in the region. Through self help group the major agricultural developmental programs such as spices production, organic agriculture practice, introduction of new and modern agricultural inputs, and promotion of marketing are carried out.

The priority of many agencies has been to develop new groups (SHGs) create interest among members for regional development at the local level. KASAM and SAMANWITA in 1990s, OMFED in 2002 and PRADATA in late nineties have started their agricultural development programs in the villages. While KASAM and OMFED mainly concentrated on the spices production and the organic fertilizer promotion, others concentrated on the other aspects of the community development.

Local and Community Based Organisation

Local and community based organizations are the grass root agencies who are really the transmitters of new technology at the village level. These include the Self-Help groups, Mahila Mandal, Yuvak Sangh, farmers' organizations, etc. All these arrangements at the local level act like the change agents in the village agricultural scenario. The aim of the organizations is to link with the outside agencies for the smooth transfer of new inputs. Within a period of five years, a large number of NGOs have emerged to work for the community development.

Farmers' Organisation

The entire discussion will be insufficient if the role of farmers' organization is not focused. In the new agricultural scenario, the role of farmers' organization is very important. Though there is no formal organization, in each village there are some units created by the people to discuss the issues regarding agriculture and farming. These organizations take the decision of not allowing people to leave their cattle free in the agricultural zone, initiative towards organic practice, issues and constraints of different public and private agencies, etc. Every farmer participates in the meeting which is organized once in a quarter or even in month during the onset of khariff and rabi season. All the farmers are invited by ward member or the influential person of the village to a particular place for the discussion. They also develop the plan outlay which is taken to the district agencies for their early actions.

II

Understanding the effectiveness of the public and private agencies is the most important part of the discussion in the chapter. As discussed in the previous chapter, there has been remarkable change in the agricultural practices in the region. There is a gradual shift from traditional rice and turmeric cultivation to modern cash crops production. Paddy and turmeric continue to be cultivated in the region. But these are cultivated in different forms by some farmers. Improved varieties of seeds are introduced in order to have better yield. New mechanical inputs such as diesel pump for irrigation, iron made plough machine, trencher and spray machines are used by some of the households. Use of pesticides, insecticide, and application of fertilizer has become necessary to some of the households. In sum, for bringing these changes the role of public agencies is crucial. There are some committed workers whose contribution is significant in the region. In this part there is some reflection on the role and contribution of public agents.

THE GRASS ROOT LEVEL WORKERS, COMMITTED VAWS AND VLWS

Since the inception of green revolution in India, there has been effort by the government to promote modernization in agriculture. In Orissa, particularly, in the rural level, the VAWs have been playing major role to popularize mechanization. They have been working as change agents in the rural agricultural scenario. They are working as the mediators between the farmers and the officials in the region.

> *The farmers appreciate the grass root workers striving for the agricultural development. They give all the credit to the grass root level workers such as village agricultural workers (VAWs), village level workers (VLWs) and voluntary workers functioning in the villages.*

COMMITTED VAWs

According to both the Kandhas and the Panas agricultural workers and officials, the VAWs are the real transmitters of agricultural technology. Because, they are primary workers and establish rapport easily with the farmers and the farmers interact freely with them than any other. The VAWs are the local members and well versed in local language. The tribals often feel free to reveal their problems to the VAWs thinking that he is a part of their society. The Farmers enquire about the new provisions available for them. The farmers pointed out that whatever the modern technologies in agriculture used today; a majority of them are due to the efforts of the VAWs than any other agricultural workers. It is in 1970s the VAWs started visiting the villages and persuade the tribals to adopt new technology. They frequently visited to the region and covered both the villages equally. VAW could succeed easily to promote the awareness about the application of modern technology among the farmers. As he hails from the same region, it became easy for him to understand the local situation. According to the local farmers VAW realizes the real difficulties of farmers. He understands the socio-economic and cultural background of the community as well as the farmers concerned. According to the farmers in the villages, the VAW tried to convince the Kandhas, who were fully tradition bound. The VAW provided the practical knowledge on how to use HYV seeds, fertilizers, pesticides, and insecticides in the agricultural field. He gave detailed instruction about the cultivation of cash crops and others in the villages. His understanding about the tribal culture and importance of scientific knowledge in agricultural field helped in disseminating technology and bringing changes in agricultural scenario. The people in the villages still remember the days when the VAWs used to come and motivate the farmers in the region. As such, people were also very free to interact with him and clarifying the doubts regarding introduction of new technology from the VAWs. The VAWs could understand the local *kui* language and motivate the people. Some said that it was of the view of the Kandhas that the use fertiliser and HYV seeds is nothing but to invite the wrath of 'Dharani Penu' and misfortune to villages. The response to new technology was less. That time it was so difficult for the VAWs to change the belief of these local people. He was initially being opposed by the community not to promote this activity and put the tribal into troubles. But the VAWs have never been harassed rather being committed to work hard to motivate the people. Being localites, they fully understand the needs of the individuals as far as transfer of technology is concerned. They understand the culture of the tribals and try to introduce the technology by convincing them. They know that the forceful implementation can produce no result.

Despite all these, the VAWs have not been encouraged by the officials for their commitment. The VAWs are often pressurized by the higher officials such as Block Development Officer (BDO) and senior agricultural officers to forcibly introduce certain technology, which are not suitable to the tribals and nor to their culture and environment. It is because traditional tribal practices were considered as obsolete and unproductive, where as modern practices were considered as superior by the higher officials. It is again, the officers have to show the achievement of targets to their higher officials by the end of the respective seasons. However, the major concern today is the substantial reduction in the number of VAWs in the block over a period of time. In 1970s the total number of VAWs in the G. Udayagiri block was 12. However, in due course of time their number has been reduced to two. The state government has not given adequate attention on the reduction in number of these grass root level employees.

THE VLWs

The VLWs are the part of the block staff. They deal with socio-economic development activities in the villages. A VLW is a grass root level worker to deal with village development work. They channellise the funds for construction of pond, check dam, village road, etc. They also work as surveyors, selectors of the BPL and APL members, distributors of credit and loans to the BPL cardholders, and selection of other beneficiaries of Widow Pension (WP), Old-Age-Pension (OAP), etc in the assigned villages. In addition, his workload is multiplied by the additional responsibility as Panchayat executive officer. He directly deals with the Panchayat members on all the village matters, which include the agricultural development too. Though he is not directly involved in the agricultural development of the tribals, his role is crucial in different ways. His commitment to serve the community is locally appreciated.

AGRICULTURAL OFFICIALS, RESEARCHERS AND EXTENSION OFFICERS

Since early 1980s the agricultural officials, researchers and extension officers have been working in the region. They have been rendering services for the growth and development of agriculture in the region. The purpose to set up the research institutes and the appointment of the technical experts is for the identification of suitable technologies and their effective dissemination to the farmers' level.

The efforts are put to focus on the development of the agro-based technologies and extension approaches. The farmers in the region are now provided with consultation based training programs on mushroom cultivation, poultry farming, use of fertilisers and chemicals on the crops, etc. Despite all this, the programs have been less successful because the training programmes which are mainly confined to the place where the centre is located. The poor people from the villages are unable to visit the centre due to time and cost factors. Many people in the villages feel that attending the programs at the centre in G. Udayagiri is expensive. The transport and food expenses go beyond their income of one day. Attending more than one day training programme is also again a difficult task for many. Many also find it difficult to go away from village during agricultural season, where people engage themselves in agricultural activities. Attending program during agricultural season is difficult for the farmers. Many also feel uncomfortable, staying away from home to attend the program. It is suggested by the respondents that it could be better if such village based training programs are organized in the village or panchayat level than at tehsil level. They feel that it will be convenient for many people to join the programs if it is organized in the villages. However, there are no such instances of conducting the program in the villages.

It is also found that out of a few people who are interested to attend the program some are not even able to attend because information on the training programmes does not reach them in time. Between the two villages, Laburi and Sudhipara, the Sudhipara people are more aware about the training programs and keep collecting update information. But in Laburi village people are aware about the training schedule. According to majority farmers in Laburi, agriculture officials rarely visit to their villages. They say that they need the visit of officials into their village. They feel their consultation on crop raising and technology adoption will help them for further improvement in practice. However, officials hardly visit Laburi. But in Sudhipara village the visit of the officials is frequent despite having some variation on the frequency among different sections of the officials (Table 5.2). The officials prefer to make a visit to roadside villages and to a few villages which are easier to cover up.

Table 5.2: Regularity of visit by various agricultural officials and the utility of the visit in the villages

Visit by the Officials	*How many times in a year on average*	*What they provide*	*Whom they focus mainly*	*Utility of the visit according to farmers*
ADAO	0-1	Just giving suggestions	Whom so ever they meet in the village	Below average
KVK	4-5	Supply of agricultural inputs like fertilizer, seeds. And other related things	Targeted people like interested farmers but confined to Sudhipara village	Medium
BLOCK AEO	0-1	Only visit	No particular target	Below average
VLW & VAW	0-1	Seeds, fertilizer	Very friendly irrespective of caste, tribe and religion	High
Others	0-1	consultation		Average

It is observed that the persons who come forward and meet officials usually take advantages of accessing new inputs that come to KVK for supplying to the poor and backward farmers. But there are many poor and illiterate farmers in the villages who are not able to access the benefits. They feel that it is not possible for them to interact with the officials unlike the educated farmers. Agricultural officials probably won't respond to them.

This also hinders their interest to seek necessary support. It is realized that if the extension workers and other agricultural officials interact with these poor farmers and ask their problems pertaining to farming and deliver service, then there will be equity in distribution and delivery of services.

Table 5.2 is figured out by taking the views of 12 members (6 from each village) with the help of focused group discussion. The farmers realization is understood on the basis of scoring out of 10 point scale, where below 3 points is considered as below average, 4-6 average, 6-8 high, and above 8 is very high.

The officials visiting the village are from different offices. They are ranging from ADAO, KVK, Block and others. But how far their visits do serve the purpose is important in discussion.

The Additional District Agricultural Officials visit the villages rarely. Their visit does not focus much on the poor and troubled farmers who often face the difficulties of crop loss and others. Their visit mainly confines to giving the suggestion only. The people who expect much from these officials feel unhappy. In comparison, the KVK officials visit the villages more regularly than the ADAO, officials. They mainly visit to Sudhipara village and target a few beneficiaries. So most of the time these farmers get benefit by receiving the agricultural inputs which are supplied to them free of cost on trial basis. But more than half of the farmers in the village are not happy because they have not been provided with such facilities.

It is noticed that between the two villages, Sudhipara people have also demand for the agricultural inputs at the offices. Their interaction with the officials including with the KVK officials is more frequent than the Laburi people. This may be the reason that Sudhipara people have linkages with local political leaders. Ramahari Pradhan a person has political influence being a BJP youth president in the district. Majority of the people in the village are his follower. So their influence in the region is also high. It is one of the reasons for the people in the villages to have better accessibility in the region. On the other hand the Laburi people are followers of Congress but there is no influential local leader of congress in the village. This way Laburi people feel that due to the difference in political affiliation this village is not able to have similar provision like Sudhipara village. During last 8-9 years the BJP party is very strong in the region. So the BJP supporters get the advantage of this for getting lion share of public provisions.

Again the visit by the AEO is also less and does not benefit the farmers as such. However, one important thing that can be observed from these villages is the visit by the VLW and VAW is also very less now days. According to the people, due to reduction in the number of VAWs, their visits have come down to one to two times a year, but it helps a lot to the people. It is not because he supplies the implements but he is friendly and mingles with all the members irrespective of caste and religion. People from the both the villages are impressed by the friendly behaviour of the VAWs. According to the people, the role of KVK officials and the VAW is important to the village. At least they visit and do some work for some people. However, the KVK officials need to cover all the villages and there should not be bias in selection of beneficiaries and giving consultation.

Despite all these issues, the visit and consultation of the officials have some positive impact on the farmers. All farmers do not consult and interact with the officials; however, a few people who often interact with the officials get the knowledge inputs and use these for the practice. The other people learn from the trained farmers and take up the new practice.

In Sudhipara village more than 15 households have received inputs of some kind from the KVK during past five years. One household in the village has received bricks and cements for constructing a tank of 4 by 8 feet size for preparing vermin compost, 10 households have received high yielding varieties of paddy, wheat, and *mungs*. A few households have received oil seeds, etc from the KVK. Similarly, a few educated farmers have consulted the agricultural specialists about the technical procedures required for selecting and raising crops.

Suryanarayan Pradhan is a thirty-three year old youth from Sudhipara village. He has been a known face among the agricultural officials for his frequent visit to KVK in G. Udayagiri. His monthly visit to the Krishi Vigyan Kendra and seeking new ideas from the officials has enabled him to become a successful farmer. There are also a few persons like him in the village who consult the officials and attend the training programs. Radha Krishna Pradhan is another farmer from Sudhipara who is regular in attending training programs. He attends the training programs and applies it on the field for growing potato and other vegetable crops. In 2002-03 he attended a district level training program. In the surrounding villages, he is now one of the influential farmers. He cultivates potato, turmeric, cabbage, cauliflower, etc. The production is sufficient for his family. He also earns better cash of approx Rs12,000 per annum from his vegetable cultivation. All his children are having school education. Out of the three children one has passed out of 9th class, two others are school going. According to him, he wants to learn more for further improvement of his practice in agriculture. He is the only male worker in his house. But he is much supported by his wife and elder daughter in his work.

APPROACH ON EXTENSION OF SERVICES IN AGRICULTURE

Over the years, the approach followed by the extension workers is from the laboratory to land. This means technology that is developed at KVK or OUAT is not by consultation with local farmers. Suddenly introducing this technology in the villages does not suit many farmers. For example, IR-1009 is a high yielding variety of paddy, developed by Central Rice Research Institute (CRRI), Cuttack, introduced in the village. Though it is widely popular through out Orissa, people did not prefer. They thought it is a long duration paddy crop and needs to be grown in low land. A few farmers experimented with this variety and failed. The land mainly being uplands and sloppy in the villages it was not suitable to grow this paddy. In this regard, Ramachandra Mallick, a farmer of Sudhipara village pointed out that he had cultivated a patch of land for growing IR-1009 variety in his field. This was a long term variety. As suggested by the officials, the low land was selected for the crop. The crops grew above the average height. But the sudden heavy rainfall during the growth stage of crops resulted in its loss and the yielding was very less. He pointed out that all his effort was gone in vain. He could have produced better had he opted other local variety or HYV paddy.

The needs of the farmers at this juncture is to protect and improve the yield of the crops they are adopting instead of promoting new crops. At present, the attention of the researchers and extension workers is to introduce new crops and new varieties but less attention to the existing crops. Ginger which was a practice of generations no more exists in the villages. The concerned agricultural experts have been failed to revive it. But there is effort from them to introduce non-local variety of turmeric and other seeds without an effort to protect the traditional varieties. As observed, the focus of extension workers has been confined to vegetable crops than on cereal crops.

BEHAVIOUR OF THE EXTENSION OFFICERS AND THEIR ATTITUDES

Though difficult, it is necessary to understand the bahaviour of researchers, extension officers and the officials associated with the agricultural development programs in the region. There are some people who are really dedicated for the developmental goal in the region and some at least show impressive behaviour by their effective communication. This includes VAWs and NGO voluntary workers, working at the ground level. As far as the extension specialists and various subject specialists of the agriculture is concerned, they are comparatively less eager to work for the farmers. They might have adequate skill to render their service but they do not have the right attitude. From the discussion with the officials during field study it was found that a few officials are more or less critical about the tribal behavior and their outlook, without having adequate understanding of tribal culture and their tradition.

The farming situation in the case of the Kandha farmers is quite pitiable as far as technological dissemination is concerned. The main views of the researchers are:

> 'The tribal farmers are backward not because of their economy but of their outlook and attitudes. They are less-innovative. Their response to adoption of new technology and desire to have better economy is less. The tribals would not accept modern technology because they do not want to change their traditional customary practices and worldview. Adherence to traditional agriculture in many tribal regions is culture based. It is difficult to pursuade the tribals to accept modern technologies. To some, culture is the major barrier in accepting new methods and technologies in agriculture by the Kandhas in the Kandhamal region. They do like to produce the crop which suits their consumption pattern. They prefer the traditional crops, which would provide them sufficient gruels and consequently the liquor for their recreation'.

The above attitude of a few officials working in the regions becomes the major hindrance in the progress of the farmers in the region. In contrast, the training organizer and the soil scientist of the KVK, from their study during 1993-94, have pointed out that the farmers have favorable attitude towards application of chemical fertilizer, HYV seed, and plant protection measures having mean score 4.46, 4.24 and 4.12 respectively. Knowledge gap to the extent of 88%, 57%, 46% and 41% was observed in case of use of micronutrients, manures, fertilizer and their management. The knowledge level in the above items was 12%, 43%, 54%, and 59% respectively. Knowledge gap of 20%, 22%, 12% and 15% was observed in varieties nursery management, seed rate and optimum sowing time in vegetables (Mishra, 2003, internal report of RRTTS, G. Udayagiri).

VIEW ABOUT THE ADOPTION OF NEW INPUTS

Many of the scientists view that the farmers are the late adopters of innovations. But the studies conducted in early 1990s by the then senior agronomist show positive result. The extent of adoption of practices like seed rate, nursery management, HYV seeds of vegetables were 82%, 76%, and 68% respectively, where as adoption of intercultural operation, manures and fertilizers and improved implements were in order of 42%, 34% and 22% respectively (Internal report, RRTTS and KVK). After interaction with the farmers it is found that these days both the Kandhas and Panas are no different from each other or from other categories while adopting an innovation. Like any others they adopt if it suits them. This is substantiated from the trend in agricultural practices and adoption of new crops, tools, and techniques in recent years.

TREND IN EXTENSION OF SERVICES IN AGRICULTURE

In recent years, the on-farm trial of crop production technology is being introduced by the researchers and scientists in the region before disseminating certain technology in agriculture. National Agricultural Technology Project (NATP) is under operation in the district since 2004. In 2005, the IMAGE, an agricultural extension institute organized a workshop to provide training support to the agronomist for the effective transfer of the technology. This workshop was organized under the National Agricultural Technology Project of the MANAGE institute in Hyderabad. Some of the scientists from the regional research station attended the workshop. The scientists were addressed to follow farming system research (FSR) approach by which the farmers can participate in the effective implementation of the technology. During their trial visit to the Katingia village on the training program it was found that there was overwhelming response from the community of farmers.

WORK CULTURE

The major problem in the region is frequent change in the staff and absence of officials in RRTTS, KVK or ADAO office at G. Udayagiri. As the institution is located in the tribal region, the higher officials posted here do not feel comfortable to spend much time due to extreme climate in the winter, lack of communication facilities and backwardness of the district.

According to the local people, each year or in alternative year, the researchers try to get transferred from the region. The spread of malaria in the region is another cause of fear among the officials coming from urban areas. From the study it is found that in a period of ten years there have been a couple of Associate Directors of Research (ADRs) posted in the region to head the responsibility. But, most of the time, the post has been vacant due to unwillingness of the scientists to stay in the region. Not many show the interest to work for the tribals and for their improvement in the region being a backward and hilly region in the state. If somebody is compelled to work in the region, he tries to take leave more frequently. So without supervision presence of higher authority, the lower officials work on their own interest. They may come to office or not, nobody monitors them. There is hardly any invigilating done, as KVK and RRTTS are the autonomous bodies. In contrast to this, there was an ADR, Dr. Manoranjan Mishra, who was committed and contributed a lot to the centre and to the people in the region. He published research papers, training manuals, leaflets about agricultural processes and crop development for the farmers. He visited the villages with his own interest and tried to understand the issues of the farmers. He preferred on farm trial by visiting from village to village. He tried to put utmost effort to remove the constraints pertaining to agriculture.

There is another institution called Additional District Agriculture Office (ADAO) located in the same block to advise and provide necessary instructions and guidelines for the improvement of agriculture in the region. The office is running with two parallel officers such as plant protection officer and the agronomist. However, many people are not aware about these two officials as they rarely visit to the field. The farmers know that the ADAO office is meant for supplying agricultural implements rather than dissemination of knowledge and techniques. Once or twice a year a very few farmers from the near by areas visit the regional agricultural centre for enquiring about the supply of inputs. However, the people from other Panchayats like Gresingia, Talarimaha, etc., and very few from

Katingia panchayat, visit the ADAO branch for the agricultural service. This shows the limitation of public service delivery in the agricultural development.

DISPARITY IN SERVICE DELIVERY

The major problem in today's scenario is the favouritism to a particular section and group in the region. Though the extension officers say that there is no such disparity, but according to some community members, the extension officials visit the villages where there is political or religious influence from the community or group. The case of Sudhipara village is an example it. Their influence in the region is one of the important factors by which the community receives lion share of public services. As far as religious influence is concerned, this is as such not clear how important it is in disparity in equity of delivery of services but this view of the local people can not be ignored.

Case: According to Gabiraju Pradhan of Laburi village, "The officials have never visited our village. We have heard that there are a number of agricultural officials working in the region for promotion of agricultural development. But where are they? We face many problems in adoption of new crops for the production process. We learnt about beans (French beans) cultivation from the people in Katingia region in mid nineties. Though the beans cultivation was started 30 years back in the Sudhipara village, our people could not adopt it till 1990s, as there was no effort to make the people understand about the adoption of the new crop. There has been no visit by officials to our village. There is no one to give us the technical support. We are not provided facilities may be because we are Christians and the officials concerned are Hindus. They may also not prefer us because we are in the remote area. Our voice is not heard, perhaps, because we are not much influential like the Kandhas in Sudhipara village. We know that they are coming to the Sudhipara village and whenever, we request them they say that we will be coming latter, we do not know sir, why they are not visiting us. We are recently cultivating new crops. Some crops are easily affected by the diseases, but there is no effort from the plant protection officer posted in the region."

On the other hand, in Sudhipara village even though there are frequent visits by agricultural officials, some people are not happy about the visit of agricultural officials. According to Ramachandra Mallick in Sudhipara village:

> "It is true that the officials visit our village. But, their visit does not make any sense. And, it creates further problem in the region. They will supply the inputs whenever they want. Even if it is off-season, they we will force us to sow the crop which, is not suitable in the ongoing season. See, it is January now; the extension officer has given us mung variety to sow in the field. How can we go for it? It is already 3-4 months late. Similarly, they will be giving us 1009 variety in September after the sowing season is gone. So there is no value of their visit. They will come and pressurize us. I do not know why do they do so? We know that they are only supporting those farmers who are talkative and followers. The poor farmers are never informed about the opportunities available for them."

During the fieldwork, it was observed that are some problems related to the distribution of seeds and other incentives to the farmers. The KVK officials visited to the village three to four times. They directly came to some houses. Upon enquiring further it was found that these houses are the regular receivers of the support extended by the officials. They were being provided with the paddy seeds, wheat seeds, new variety of vegetables, etc. Similarly, instructions and guidelines were given on how to construct the place for storing cow-dung, and other waste materials for preparing compost; the preparation of vermin compost, etc. But the important issue is that the supply was restricted to certain influential members. The non influential, poor groups, and illiterate farmers participation was less. They were neither informed nor provided with any support for their improvement in agriculture.

TECHNOLOGY TRANSFER AND FARMERS PARTICIPATION

Over the years the approach for technology dissemination has been training-and-visit (T&V). Particularly in the regional research centre the same model is being followed. As a result, the farmers' participation has been restricted without any substantial progress. On farm trial is done only once during the introduction of new technologies or inputs. Many a time, the officials distribute the seeds freely, but to certain section of individuals. There was time when free seeds, fertilizers and other inputs were supplied when the demand was limited. People were ignorant about the new technology. The involvement of VAWs and the voluntary workers supported the cultivation of new crops in latter period. Now, they need much support from both the government and the private sectors. They require updated technical support for agricultural development. They are eager to use the modern tools as well. However, it has not been possible due to the fact that the extension approach of service provider is not favourable to many. There are certain crops where the farmers do want to follow traditional procedures. At this juncture it is essential to consult the farmers on what do they really need. However, their need is not addressed with utmost importance. Whatever limited inputs are supplied to limited farmers also do not work out without interest of the farmers. For example, the supply of HYV paddy IR-1009, wheat seed and hybrid turmeric seed do not generate interest among the farmers. Therefore, the supply of these seeds becomes waste. Some people pointed out that limited farmers who receive these seeds for trial often consume instead of using them in the field.

Similarly, very few people collect chemical fertiliser from KVK free of cost. They are supplied to use in paddy field but some of them use in vegetable field rather than using in paddy field. Because the farmers feel that if the chemical fertilizer is used in the field, land would consequently require more application of the fertilizers in the following years. It is really difficult for the poor farmers to use chemical fertilizer by purchasing at heavy price. In addition, the use of fertiliser in vegetable field becomes more essential than paddy field.

The farmers interested in adopting those innovations, which were favourable for their climate and did not bring any harm to their environment. The Kandhas realized that adoption of new varieties of seed in paddy such as Jajati and Lalat are favorable to their climate and won't create any harm. Therefore, it was easily accepted. However, some other varieties like IR-1009, IR-1014, Swarna Masuri, Masuri, etc did not suit to their climate. After several trial and error instances they were rejected. However, the pressure from the extension officers creates problem among the farmers. They force the farmers even without knowing their interests.

According to some of the respondents in the village, the officials' impositions of adoption of hybrid varieties are responsible for the loss of their valuable traditional crops. Since 1980s, the farmers are not able to protect their traditional crop, which possess rich nutritional value. The people have lost their traditional occupation of ginger cultivation either due to the change in climatic condition or due the loss of qualities like taste and production quality of traditional seeds. According to the director of the NGO, 'ASSART':

> "the ginger cultivation was stopped due to the entry of the new varieties imported from Kerala. The farmers were given false assurance that the cultivation of new ginger types would give better productivity, better aroma and would be more resistant to the diseases than their traditional types. They were supplied with the new ginger seeds through the spices board that is located at G. Udayagiri block. However, that was not effective in the villages. Rather, the introduction of the new varieties resulted the loss of original ginger crop. The farmers lost their traditional variety of the ginger. Now, there is no clear differentiation which one is local and which one is hybrid variety."

Several times, the farmers have requested the technical experts and the specialists regarding the above issues. But the issue still persists in the region. They have been given assurance to supply the necessary inputs for their survival, but that has not materialised so far. The procedures which have been suggested are not successful. In this regard, many scientists of the local agricultural research centres claim that despite frequent instructions on seed treatment, mulching and others given to them, to follow in ginger cultivation, the tribal farmers do not adopt the necessary technology nor follow appropriate mechanism like proper seed treatment, soil preparation and chemical pesticides application, etc in order to protect the varieties. But, the Kandha farmers view that the procedure has been followed according to the guidelines of the researcher. But, no good result has been obtained. They have been trying to follow the seed treatment as per the guidelines. Many of the farmers pointed out that the chemical pesticides, insecticides and fertilizers suggested by the scientists are not available in the market nearer to them. These are available in markets like Bhanjanagar, and Berhampur, but not easy to access by a poor farmer because of many reasons like expensive travel-cost, time required, etc.

The other important issue as raised by many farmers is that the insecticides and the pesticides suggested by the pant protection officer are less effective to protect the plants from getting affected. Despite that, the Plant Protection Officer working in the region is less concerned about it. At present, there are two agronomists working in KVK and office of the ADAO, one in each institution. Though the agronomist in KVK is often enthusiastic to work for betterment of the farmers, he faces problems with shortage of staff. It is observed that the farmers have been suggested to follow certain procedures to protect their crops, but no practical action has been taken in this regard. No scientist has ever tried to provide the instruction on the spot. No sample of the affected areas has been taken into consideration. In contrast, the scientists claim that they are following appropriate technology but the failure still continues. According to them, "if the farmers would follow wrong procedures, how the crop can grow well".

But the farmers pointed out that ginger and turmeric are their traditional crops so they know the procedure of cultivation. They had been doing for more than five decades. Why suddenly after 1980s the ginger crop could not grow well? The seeds get rotten before germinated. If they are not following the real procedures how it could succeed in earlier period is the question the farmers ask.

ROLE OF PRIVATE AGENCIES AND ITS EFFECTIVENESS

As far as turmeric cultivation is concerned, the traditional variety of turmeric was replaced by so-called HYV/improved variety of seeds in some areas of the region. It is felt that this has affected the loss of gene pool of the traditional variety. With the addition of the genetic modified crop for the sake of improved production, the tribal lost the quality of the traditional variety of turmeric and ginger. It is now a grave concern among the tribal farmers, who have lost the ginger cultivation and gradually have experienced the loss of good productivity of the traditional turmeric variety. In 1980s, the KASAM, an agency, promoted the use of HYV varieties and application of chemical fertilizers in the region. In Katingia and in nearby areas, the people used the varieties supplied by the agency. However, that couldn't be successful may be of the reason that it needed better seed treatment than the local ones. The water requirement was more. It also needed more labour input. According to the Kandhas, in 1993, the improved ginger variety that was brought into the district did not work well. It was black in colour in the middle and after a few months of storage they got rotten

In 1997, KASAM supplied five varieties of seeds such as Lagdong, Rajendra Sonya, Roma, Surama and Alipi to the farmers in the name of improved seeds, and farmers said if they grow those varieties they will be given crop insurance. All the harvested products would be collected by the KASAM with a little higher price than the existing market price. Many villages including the study villages in Katingia G.P, and other regions in Kandhamal district used the improved seeds instead of their traditional varieties in large number. But there was no substantial result observed. The production was rather less than the traditional variety. Lagdong, a hybrid variety could be produced of the same quantity like local variety, but needed much caring than the local ones. It needed raised bed plantation. Mulching was needed in the same way as it was in the local variety. Similarly, Rajendra Sonya, another variety, though it was scented one, could not attract the producers as well as the customers due to less demand for it in the local market. The other varieties such as Surama, Roma, Kasturi, etc. were more scented, but yielding from these was less. Therefore, the production of the hybrid varieties didn't attract the farmers.

SPICES DEVELOPMENT SOCIETY (SDS)

Spices Development Society has been formed in the region in 1996 to promote spices production. So far 10 SDSs have been developed in the region. The main role of SDS is to promote the produce and earn better profit in the international market. In Katingia also similar SDS was developed in 1996-97. Seeds were supplied to farmers who became the members of the society. The funds were raised with the support of IGRY. Now, KASAM has target to raise its members to 12,000 that would include both male and female with the formation of 61 societies.

OMFED

Initially, OMFED developed as a marketing agency for the dairy products. However, in due course of time it could extend its operation in the other areas such as spices production and marketing in Kandhamal region. In 2002, OMFED worked jointly with KASAM in order to enter into the spices market in Kandhamal district. It supported the KASAM in the form of financial support and the other technical staff for exploring new idea in the spices sector. Similarly, KASAM cooperated with OMFED by sharing the ideas and experience in the field of spices production and marketing. But, after two years of joint working OMFED got itself separated from KASAM and emerged as an independent organization to focus exclusively on the women who are the real farm laborers in the region. The OMFED focused on both tribal and non-tribal women to form women groups in the region. The group that was developed by OMFED was considered as 'Mahila Mandal'. In due course it emerged as a '*mahila* self-help-group'. A separate project, "Organic turmeric project for women farmers", was launched in 2004. A few women technical experts from with agricultural science background are now working for the organisation with a low salary of Rs. 4,000/month. OMFED appointed them in the region to promote *'mahila group'*.

ORGANIC TURMERIC PROJECT FOR WOMEN FARMERS

The objectives of the project are:

1. To provide appropriate technical support and instructions for turmeric production in organic ways and to increase the quality of the produce.
2. To supply improved varieties of seed, bio-fertilizer, pesticide and other supports for the production of turmeric and other spices items in organic process.
3. To supply the necessary inputs like compost, vermin compost, bio-culture, etc.
4. And, to create awareness among the women for micro-saving towards self-dependence.

In 2004, OMFED supplied a few varieties of turmeric to the women farmers through *'mahila mandal'* for the production of turmeric.

CONTRACT AGREEMENTS BY THE KASAM

Like OMFED, the KASAM is also moving ahead with agreement paper. The agreement paper is being signed by the farmers in view of market reforms. The contract farming is being promoted with the participation of private agencies. As a result, the farmers or the landholder would produce the crop as designed by the agency. The agency will provide the technical input as required for the crop production after the agreement. The farmer or the cultivator has to obey the guidelines for the organic cultivation. He or she has to strictly abide by the agreement that he wouldn't use chemical fertilizer, pesticide or insecticide in his field. In addition, he has to use only the organic seed for sowing. The harvested produce will only be sold to the KASAM, if KASAM is not ready to buy, then only a farmer is allowed to sell outside the agency.

Over the years efforts have been made to develop the district into an organic zone by developing interest among the farmers to opt for organic processes than the chemical ones. The NGOs, KASAM and SAMANWITA are the two agencies working for the promotion of organic cultivation. The initiative taken by the KASAM in the zone is not very old. Similarly, the OMFED as an autonomous marketing agency in the government of Orissa has objectives for the betterment of people. But similar conditions and agreements do not suit many people in the region for the following reasons. The women members have to sign an agreement to be associated with the group for both production and marketing. The objectives are to develop the awareness and interest among the women farmers in the Kandhamal region. But, the eligibility criteria asked for being a member does not suit many poor women to be the member of the community for getting the membership. A woman must have land under her ownership. She must sell the produce at the society. She shouldn't involve in any agency other than OMFED for turmeric or spices business and she has to submit a written agreement to the society for supplying the turmeric to the society.

FARMERS' OPINION

Over the years both private and public agencies have been taking interest to promote the improved varieties of seeds for the turmeric and ginger cultivation. However, this instead of creating interest among the farmers has raised tensions among the farmers. People claim that the new artificial turmeric varieties are not up to the quality. The quality of the improved varieties is better in aroma and colour but with regard to the disease resistance, the hybrid variety is far behind the local varieties. The private varieties require the application of fertilizer, raised bed planting, timely seed treatment, proper storing, etc. On the other hand, the local varieties do not require any care. The cultivation is done without the application of any fertilizer. The quantity of production from the local seed is usually more than the so called improved variety. The marginal cost of production does not affect any group of farmers for the operation. The most important thing is the sustainability with the local variety. As there is no application of chemical fertilizer, pesticides or herbicide, the production is sustainable than the production of the hybrid varieties. This is the reason why people use the traditional variety for the production of the turmeric and ginger. However, without understanding the people' interest in the production of traditional crop variety, the efforts have been made to change the cropping pattern in their traditional operation.

What they Need

There are certain sectors where the farmers need the involvement of the agencies. The farmers in the region have better knowledge in agriculture. They know the process and techniques of cultivation and production. But there are certain areas, where there is the need of the extension agents' involvement. As already discussed, the most important issue in the area of agriculture is lack of irrigation facilities, but the other recent problem where the people confront is control of pests and insects. It is observed that the diseases like curliness of leaves, white and burn disease of bean and potato leaves, and rottening of ginger are common. Therefore, there is urgent need of the agricultural specialists to tackle the problems.

IRRIGATION

There was a watershed (Kamasuga) project developed in 1995-96 at Budedpara. Integrated tribal development agency (ITDA), Baliguda supported the water harvesting system at Tiangia in the region in 1985-86 and two WHSs at Katingia and in 1990-91 at Katingia and Lingagada, each having one. In 1970s, the ITDA started the construction of dug-wells in the Sudhipara village. So far, there are 3 dug-wells constructed by the ITDA in the Sudhipara village and one in Laburi village. There are seven dug-wells in Sudhipara village and two dug-wells in Laburi village. Some of the dug-wells are developed with the loans sanctioned by the IRDP scheme through District Rural Development Agency (DRDA).

Despite some initiatives in the region to promote irrigation for the perennial crop production, the efforts have not been successful. Many a time, the farmers face difficulties to receive the required water for the rabi crops. During this time, the springs dry up. Therefore they either depend on the well or the *chuas* in these areas.

During mid nineties, an NGO, PRADATA had taken the initiative to develop watershed development program in Tiangia region of the Katingia Panchayat. However, it could not succeed. The NGO could complete it's work. After a few years gap, the DRDA and soil conservation department with the help of irrigation department planned to complete the work. That project has helped some people in the region.

In sum, the post independence period envisaged the growth of multiple number of public and private agencies both in urban and rural areas. The institutional involvement has been crucial in agricultural development both in the farm technology transfer and other service deliveries. But the equity and distribution of services is a challenge in many regions. The Kandhamal district is too not spared from this challenging situation. The differences in delivery of services are noticed even between two villages located nearby. So what matters is the behaviour and attitude, willingness and interest, biasness and seriousness of the officials on one hand and the influence, willingness and interest and attitude and seriousness of farmers on the other hand which play crucial role. Not all the farmers or officials do have same seriousness or willingness. Some of the officials like VAWs or a few extension workers are more serious in their work. Some NGOs do really needful for the farmers as in the case of PRADATA. Some take serious note to provide utmost service for the good cause of farmers. But still there is something lacking. Less effort has been put to understand what really the farmers need. It is not one man's job to come forward. Combined effort should be made and action should be taken as per the needs of the farmers.

❑ ❑ ❑ ❑ ❑

CHAPTER - VI

NEW ECONOMIC POLICIES AND THEIR IMPLICATIONS FOR AGRICULTURE

INTRODUCTION

Introduction of economic reforms in June 1991 is a landmark in the economic policies in India. This new economic policies came to the fore with reduction of government's control on various aspects in domestic economy. In other words, this enhanced involvement of private companies or agencies and reduced the government control in important sectors. Liberalisation reduced the tariffs and removed the major barriers for the entry of multinational companies to invest in India. Liberalisation, privatization and globalization, which are the three major components of the new economic policies, came into forefront since early 1990s. Though the economic reform measures are formulated keeping the security and protection of vast mass of agricultural community into consideration, there are many instances in India where the farmers are affected by the new economic policies. The case of Kasipur in Rayagada district in 2001-02 and Kalinga Nagar in Jajapur district in 2005 in Orissa can be cited here. There are many other cases in Orissa where the tribal people and the rural masses have suffered due to the effects of industrialization and privatization. The proposed special economic zones (SEZ) are a byproduct of the new economic policies in recent years. The cases of Singur and Nandigram in West Bengal in 2006-07 where the agricultural lands acquired by private industries such as TATA group in India and Salim group of Indonesia have raised the conflicts between the farmers and the state governments. But there are also cases where the entry of companies and industries is linked with agricultural purposes. The land is acquired by the corporate giants such as PEPSICO, Reliance industries, etc for farming. This is called corporate farming. The case of contract farming is slightly different from corporate farming. In this form of farming the ownership on land remains in the hand of farmers. But the farmers are regulated by the companies both on crop selection and market control. The case of contract farming, though not new in India, is now a cause of worry in many parts of the country. New economic policies give more scope to the private agencies and multinational companies (MNCs) to undertake cash crop production by contracting the farmers. Similarly, the liberalization concept has become apparent in many sectors, particularly in non farm sectors. However, the farm sector has been equally affected by the liberalization process. Say for example, the liberalization has removed the inter-regional and regional barriers in the export and import of agricultural products ahead of new economic policy. The economic reform per se has emerged in different forms in India. The different facets of the economic reforms through new economic policies have generated new problems in agricultural sector both in rural and tribal areas.

INSTITUTIONAL REFORMS AND AGRICULTURAL DEVELOPMENT

Credit Reforms

At the institutional and sectoral level, the rural retrogression could be seen in terms of regress in several aspects that are crucial for small-peasant agriculture and rural economy.

After a surge following certain radical institutional and structural reforms in the late sixties and early seventies, there has been a drastic decline in the share of institutional credit to agriculture. State investment has come down not only in the infrastructure like irrigation, but also in critical areas of research towards development, multiplication and distribution of new varieties of seeds and new agronomic practices. The entire system of agricultural extension has been jeopardized and starved of staff and resources. A study of credit from formal institutional sources shows that between 1980-81 and 1999-2000, share of agricultural sector on short-term credit declined from 16.9 per cent to 8.3 per cent in India (PSM. Rao, 2003). The acceleration in the decline in share of much needed long-term credit for investment was witnessed since early 1990s. Beginning with the early 1990s especially since 1993, the small borrowers' share in bank credit declined steeply from 21.9 percent in 1992 to 7 percent in 2001(ibid.)[1].

The impact of credit reform has a direct effect on agriculture. As most of the farmers, particularly from backward areas rely on the formal institutions and depend on credit support for their agricultural operation, the credit policy is a very crucial matter. Any reduction on subsidies and supply of credit thus directly influence the farmers and the agriculture as well. Since independence till late nineties, there was individual credit to the farmers for their agricultural development. Since the emergence of economic reforms there is certain change in the institutional activity by emphasizing on supply of group loans than individual loans to the people in rural areas. As found from the fieldwork, 59 households from Sudhipara village have availed credit facilities through SBI, Katingia branch from 1993 to 2003. The loans have covered horticultural development, plantation, goat rearing, livestock purchasing, etc. More than 50% of households had received subsidies of 50 per cent of the total loan sanctioned. Very interestingly, in the villages, most of the beneficiaries have received the credit from the year 2000 to 2004. But in most of the cases the credit type is short-term. Further, in 2004-05 the government through cooperatives started emphasizing on supply of the group loans through self-help groups. Though the self help groups emerged much before 2005 i.e., in 1991-92 in the region, they have become active recently. The effect of reduction of individual loans has not immediately shown any negative impact. However, in long term, it will lead to negative consequence on the agricultural development in the region. The small farmers who urgently need loans have to depend on the self help groups. When the farmers are not able to repay the interest on their borrowed credit at a very low interest rate, how can they be paying back the loan with exorbitant interest rate charged by the influential self help group members! The self help groups have been active in the region in distributing credit. They are charging Rs. 5-10% interest rate per month, which is very high. The clients are mainly the small tenants and marginal farmers, who are unable to get loan from the bank. It is observed that some people have been considered as defaulters by bank for their failure to repay the borrowed money even after frequent warnings. They have been denied to access any loan further.

CASE-6.1

Driti Krishna Mallik of Sudhipara village is one of the farmers who have been considered as defaulters by the SBI Katingia branch. He had borrowed loan of Rs 25000 with 50% subsidy for investment in vegetable cultivation on 28th Mar 2000. The date for loan repayment was on 25th Feb, 2003 which is after three years of sanctioned date. However, his investment plan did not become successful. His mother became sick then followed his other family members. He spent some of his money for this urgent expenditure. He frankly pointed out that he could not invest all the borrowed money for vegetable growing. He has spent a chunk of the share of borrowed money for urgent non agricultural purpose. But after he became failure it has not become possible for him to access any other loan further from the bank.

There are several cases in the villages where some borrowers have been included in the list of defaulters. For these people a self-help-group is the important source to borrow money.

For the bank employees the rise of number of defaulters is a problem. There were seven cases of defaulters in the villages during the time of fieldwork. It is found that at the time when there is increase in trend towards new crop production, technology adoption and mechanization there is also rise in number of indebted who are often considered as defaulters by banks. This is mainly the case with the small and medium farmers who invest money for improvement in agriculture. But after failure of crops or due to other problems as in the case of Driti Krishna Mallick these farmers are being failure to repay the loan.

The observation in the village level self-help-groups (SHG) is that the responsible positions such as secretary, president and other related posts are occupied by the influential members who have political affiliation and have sound academic background. Their influence in the group does not allow the poor farmers to get into the group. It is also observed that the tribals who are not fluent in speaking oriya and feel shy to mingle with the officials and non tribal members do not get the reap of benefit out of the self help group venture. Thus, the effort to put the poor onto priority has become failure. The exorbitant rate of interest charged by the self help group for the loan seeker is not a good practice. The marginalized and poor farmers who borrow credit do face problems of repaying due to high rate of interest. At present, most of the self help groups such as MAA at Munda Naju village, Maha Laxmi in Sudhipara, etc are charging 5% interest per month. So this is much above the regional and nationalized banks' interest rate.

Though it is often claimed that the SHGs benefit the people and empower the groups, but, by and large, the small and poor marginal farmers are sufferers. The small landholders without institutional support are being the victims of these groups. No doubt, the SHG members are benefited from their collection of interest and the loans from the borrower and exhorting pressure on the loanees. It is found that many poors, landless labourers and marginal holders are unhappy with this trend of economic activities of SHGs. But they are helpless because they have no alternatives. They feel that for this credit problem government is responsible. As many of them are not aware of credit policies and the institutional policies they accuse the government for their problems. It is felt that a SHG is not responsible for charging more interest. Every SHG wants to create funds through various means like money lending. District rural development agency pressures the NGOs and voluntary agencies to produce the progress reports about the SHGs. So there is indirect pressure from the NGOs on the SHGs to show the achievement and the status of funds. Ultimately, SHGs target the clients for increase their funds.

On the positive side, a SHG is the medium through which the entrepreneurial skills of the group members have increased. The women members have been more or less active and self empowered. But there is a need of institutional focus for the protection and development of rural poor and small landholders. Much emphasis should be given to protect these marginal groups. As agriculture is the primary occupation of majority of the people in the region, the people need financial support (loans) through easy repayment. However, the SHGs have not been helpful in this regard.

There are many instances in the village where the farmers urgently need the financial support due to sudden failure of crops, irregular rainfall or for the farmers who want to buy agricultural implements, and land development. But, they do not get timely support from the credit institutions to deal with these activities. In this regard, the bank officials say that, the increase in number of defaulters in the region worries them to release further loans. There are now more than 60% of the total borrowers who have not cleared their debt and may be enlisted in the defaulters list. So increasing number of defaulters in the region is also taken as the major hurdle for flow of finance in the region. On other side, people often claim that the purpose for which they seek loan and the time they need is not met by the credit agencies. Therefore, the loan is utilized for other purposes.

Many people wait for the time when government would support them. In 1980s the then Chief Minister of Orissa, Shri Biju Pattnaik had declared to free the tribal from credit burden. So even today some people expect their loan can be exempted from repaying. A few people, who are aware of crop insurance, demand at the regional banks and the Rural Development Agency to extend insurance support for them.

The other observation in the villages is that there are a few big farmers who often lend money. For repayment of the borrowed money, the borrowers are asked either to pay by cash or kind. In cash they pay the similar interest, i.e, 5% per month as charged by the SHGs in the region. In kind, they contribute their labour in the moneylenders' agricultural fields during agricultural season. Some borrowers give agricultural produce in the form of kind to repay their loans. They give agricultural produce to their moneylenders at the price as fixed by the moneylenders.

TAX REFORMS IN AGRICULTURE

The major changes that are observed in tax reforms during these days have created difficulties for the farmers. Particularly, the water taxes have been increased four-five times within a period of 10 years. The memo no 1108 date 27-04-2005 of G. Udayagiri tehsil has given the guidelines to all revenue offices in the tehsil to charge Rs. 450 annum per hectare of irrigated land. As discussed with the revenue inspector, the tax of Rs. 450 is charged to the land holders having irrigated land. This is charged against the paddy land. Again, Rs 280 is charged against vegetable cultivation per hacter per annum[2]. Being a fifth scheduled area and tribal populated tehsil there is no exemption of water taxes for the tribal landholders in the region. This has already created some disturbances among the tribals and the SC communities in the region. Many landholders have no proper land record even after five decades of land reforms. Therefore, it will be a great difficulty for the lower revenue officials to assess the cess to be paid by the farmers.

In view of above, the villagers of Laburi pointed out that a few acres of lands are irrigated with the support of canal constructed in Tiangia village. The construction work is still unfinished and most of the land is un-irrigated as usual during the summer and winter seasons. However, tax has been imposed to all the families who have land, be it irrigated or un-irrigated, in the villages with the claim that the land is located nearer to the canal.

REDUCTION OF SUBSIDIES ON FARM IMPLEMENTS

Reduction of subsidies on farm implements is one of major problems for small and marginal farmers in the region. There has been reduction in the subsidies on the needed farm implements like iron-plough machine, trencher, spade and other essential small farm implements. However, the subsidies are given on heavy machines and large implements like tractor, winnower, etc, the access of which is not possible for the small and marginal farmers. The letter no. 1s (15) 3/2005-710/agril.dt. 14-10-2005, Office of the Director of Agriculture and Food Production Orissa: Bhubaneswar has given the directive to the regional agricultural service centres that there should be 25% subsidy on thresher and Rs. 30,000/- subsidy on tractor. On the other hand there is no guideline on giving subsidy on garden implements like plough, trench, hoe, wrench, etc. According to Agricultural Overseer the small agricultural implements had 50% subsidy 3 years back. But now there is no subsidy given on these implements. The supply of these implements has also been reduced. Now the iron plough is sold at Rs. 300 per piece. Keeping this view the farmers realize that it is better to depend on wooden plough rather than looking for iron one. But depending on wooden plough means the loss of time. Cutting trees or collecting timber for construction of this leads to destruction of forest. In addition, it is not easy to make by all people. This needs the support of an expert who makes this. So during the crop growing season, the cost of making a plough goes high. But due to availability of timbers and the iron part at the local blacksmiths, it is often favoured to use wooden plough than using iron plough at a higher price. This is because, the cost of preparing a wooden plough reaches to Rs. 200-250 while an iron plough is now Rs. 100 more than the wooden one. In addition, the stock available at the regional agricultural centre is very less. A farmer has to visit several times to get this, which is also not easy for a farmer.

The government policies by increasing subsidies on tractor and other heavy implements encourage the big farmers rather than small and marginal farmers. But in tribal and backward region, this is not as useful as majority of the farmers are poor and marginal ones. The other point is that in tribal and rural villages women play important role in agriculture. Therefore, use of tractor and other heavy instruments means removal of agricultural labourers from the agricultural occupation. Though in the village there is only one tractor, this tractor is used by the big landholders who were earlier engaging women and landless agricultural labourers in their agricultural operations. Therefore, the inference that can be drawn is that in near future with the support of subsidies on tractor, rich farmers, both tribals and non tribals, may possess tractors and directly and indirectly remove the land less agricultural labourers and women labourers from the agricultural wage labour.

PRIVATISATION OF FERTILISER SECTORS AND FARMERS' PROBLEMS

The reduction in the control of fertilizer sector has promoted the involvement of private agencies such as Mahindra and Mahindra, Nagarjuna, Paradeep Phosphate Ltd into fertiliser business. The privatisation of fertilizer sector has produced mixed result. On positive side, the farmers have got the benefits of large market opportunities for accessing different compositions of fertilizer. Urea, Shymala, Diammonium Phosphate (DAP), Potash, Super potash, etc that are available. The availability of fertiliser of different compositions has helped the advanced farmers to use appropriate fertilizer with the consultation of the specialized agricultural officers appointed in the region. In recent years, some farmers have definitely been benefited out of the production made with the application of fertilizer, particularly on the vegetable crops.

On the other hand, the negative effects of privatization are many. From the study it is observed that one of the reasons for farmers not using fertilizer in their land is reduction in subsidies which has troubled farmers to buy fertiliser. From the interaction with the farmers almost all farmers had similar remark about the application of fertilizer though they say there are other factors which the farmers are lackluster of using fertilizer.

The price has been double in some fertilisers within five years. Figure 6.1 given below shows that the cost of urea has become even more than double within five years. In 2000-01 the cost of urea per kilogram was Rs. 3.00 but in 2005-06 it is Rs. 8.00 per kilogram. Similarly, there is one and half times increase in the price of Shyamala. The concern is that the most sought after fertiliser, Urea has been costly and now it is Rs. 8.00 per kilogram. Urea is a common fertiliser used in land for both cereals and vegetable growing in the villages. Upon asking the agricultural overseer at ADAO, G.Udayagiri it was found that Orissa government has no provision of subsidies on HYV seeds. On the other hand the supply of subsidies by primary agricultural cooperative societies has been gradually stopped. Though in Raikia LAMPS there is supply of fertiliser but the sale price is same as with private shops located both in G.Udayagiri and Raikia towns in the block. Therefore, people often purchase it from the private shops on their visit to any of the places.

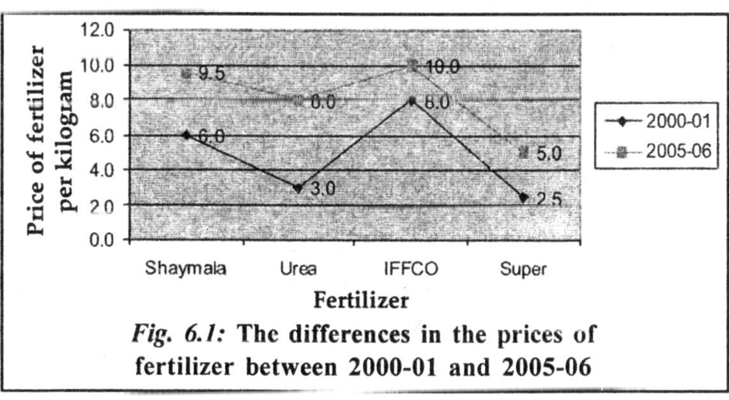

Fig. 6.1: **The differences in the prices of fertilizer between 2000-01 and 2005-06**

Source: Sudhipara farmers who use fertiliser on their crops.

From the previous chapter it is clear that there is growing interest of adopting vegetable cultivation in the region. Similarly, there is increase in demand for pesticides and insecticides. But, at present, there is no provision by any government agricultural institution for supply of these inputs in this backward district. This is true that the pesticides and insecticides are available in the local G. Udayagiri and Raikia markets. But, the cost of these is not favourable to many farmers to use on their infected crops. The common pesticides and insecticides that are used in the region are malathene, bhavistine, dethine, phurat etc. These are mainly used for the beans plants. Malathene is mainly used for cauliflower. According to the farmers this is used mainly to enable the cauliflower growing quickly. It is observed that a few people know about the value and benefits of using these pesticides. However, many people are aware about these pesticides but do not use them because of cost. According to some farmers, 100gms of malathene costs around Rs100 to Rs 120. So it is difficult for many poor farmers to think of its use. Similarly, the other pesticides and insecticides are also very costly. Therefore, it has become difficult for many to use them on their agricultural field. Though government is worried about the less productive agriculture in Kandhamal region, however, due to increase in price of fertilizer, pesticides, and insecticides, it has become difficult for large number of farmers in the district to apply these despite having knowledge of application.

The application of fertiliser, hybrid seeds, pesticides and insecticides and modern farm implements has been directly affected due to gradual reduction of subsidies over the years.

POWER SECTOR REFORMS AND IMPLICATIONS FOR AGRICULTURE

With the advent of new economic reforms, the Orissa government pioneered reforms process by enacting Orissa power sector reforms act, 1995, which came into effect on 1st April, 1996. Orissa is the first state to privatize electricity sector. The privatization of electricity sector has resulted in the establishment of Grid Corporation of Orissa (GRIDCO) into transmission and distribution functions. In Kandhamal, SOUTHCO division of GRIDCO is regulating the above activities. This has enabled the state government to gain profit. However, it has resulted in negative effects on the small and marginal farmers who are unable to pay huge service charge for their agricultural purpose. The privatization of electricity is a major hurdle as the farmers who are newly entering into the entrepreneurial farming. According to vegetable cultivators the electricity supply to the village does not benefit them as there is no regular supply of electricity. The farmers who are willing to get the connection for the electricity supply do not reap real benefits because of irregular power supply. Second, the major problem is that neither tariff nor subsidies are provided for the farmers in this tribal region. So for the new entrants into vegetable cultivation, it is an uphill task to succeed.

There was time when a person in rural area had no willingness to use lift irrigation for agriculture. There was effort to create awareness among them. The green revolution mainly emphasized on modernization of agriculture in the form of supply of hybrid seeds and development of irrigation support. The awareness of the people to use hybrid seeds, irrigation and mechanization was very less till 1980s. But, many people are now aware of mechanization and modernization despite their inability to access the new inputs. More than 90 percent of people are aware of diesel pumps, fertilizer, irrigation, chemical pesticide and insecticide, etc in the villages. By the time they are aware of, the supply sector has not been as active as was before. For watering plants, the use of electric pump has become costly as there is no subsidy.

In present agricultural situation when the irrigation has been a major requirement for effective agricultural operations, it has become necessary for many farmers to use the pumps for irrigating the vegetable fields. This is observed basically among some of the medium and rich farmers in the region who are willing to go for off-season vegetable cultivation. In off-season vegetable cultivation, irrigation support is must as the existing water sources become dry during summer. The farmers who have dug-well or small ponds thus want the government to provide subsidy support on electricity supply for the irrigation. However, there is no substantial support from the government even in these backward areas. Many poor farmers who are interested for the off-season cultivation are unable to take up this.

According to a farmer, Radhanath Pradhan, "We have electric supply in the villages. But many people are unable to go for the connection as the service charges are not affordable to poor farmers. Use of pumps for irrigation is not possible for many medium and poor farmers in the villages". This is one of the factors those hinder the interest of some farmers to take up off-season vegetable cultivation.

ROLE OF LARGE SIZE MULTI-PURPOSE COOPERATIVE SOCIETIES (LAMPCS) IN AGRICULTURE

The Phulbani LAMPCS was registered in 1976, through the en bloc amendment of the bylaws of the Phulbani Service Cooperative Society. Three other similar service cooperatives were merged into the Phulbani cooperative. It draws its membership from 215 settlements under 11 gram Panchayats. Of the total membership of 5737 individuals, 3100 belonged to scheduled tribes (mainly of Kandha tribe), and nearly 1500 were from scheduled castes.

SERVICES TO MEMBERS BY LAMPCS

The bylaws of the Phulbani LAMPCS expect it to provide credit, agricultural input supplies, essential commodities and other daily consumption items to members, and to market their agricultural and minor forest produce. The loans disbursed to members in the last few years were as shown in Table 6.1.

While loans disbursed each year were modest in terms of numbers of borrowers and amount of loans disbursed, there were 1294 members with loans outstanding in their names (on 31.3.2001). Of these, 719 were defaulters. The cooperative had not issued any medium term loans for several years.

Table 6.1: Activities undertaken by LAMPCS, Phulbani from 1999-2001

Sl. No.	*Crop Loan*	*No. of members*	*Loan in cash*	*Loan in Kind*	*Total Loan (Rs)*
1.	Kharif 1999	221	795,000	198,000	993,000
2.	Rabi 1999-2000	105	210,000	93,000	303,000
3.	Kharif 2000	259	1,036,000	193,000	1,229,000
4.	Rabi 2000-2001	96	234,000	215,000	449,000
5.	Kharif 2001	157	714,000	140,000	854,000

Source: Sashi Rajagopalan, 2002. Study on Tribal Cooperatives in India, ILO report.

Chemical fertilizer was purchased and supplied to members by the society in the form of loan in kind. The maximum supply in any year did not exceed 100 quintals. Sale of essential commodities and other daily requirements ranged from Rs. 1.8 million, to Rs. 4 million in any year. The cooperative did not procure produce from its members. Members said that they used to sell turmeric, tamarind, and mahua seed through the LAMPCS earlier. However, as the rates tended to fluctuate widely, the cooperative had stopped the procurement of MFP. For turmeric, the prices had fluctuated in the previous year between Rs. 15 and Rs. 22 a kilo, and the rate at the time of the fieldwork (2005) was Rs. 10 to 12 a kilo.

The cooperative society had constructed several go-downs and shops for its own use, but has now let out most for rent to others, some of whom have now sub-let the space, given the vulnerability of the cooperative. Large numbers of members and non-members thronged the ration shop run by the cooperative. Members said that they worked on their farms for 6 months of the year, growing turmeric, vegetables, and rice, the last, for their own consumption. The other 6 months, they collected forest produce – kendu leaf, medicinal herbs, amla, tamarind, honey, and firewood. They were not clear about to whom the cooperative belonged to, but said that it was meant to service them.

PROFITABILITY OF LAMPCS

The audit report of the cooperative indicated that the cooperative had incurred a net loss of Rs. 226,000 in 1999-2000. Audit for the next year had not been completed. Accumulated losses were around Rs. 2.5 million. Nearly Rs. 2.1 million was set aside for bad and doubtful debts, and this was the main cause for the large losses. It is likely that during the course of reorganization, the losses of some of the merged cooperatives also added to the overall losses. As the cooperative appeared to provide few services, and those, too, at rather modest levels, it appeared difficult for it to be able to be profitable for its members and itself. Of the total share capital of Rs. 760,000, Rs. 187,000 had come from the government. It was not clear whether the rest had come from members voluntarily, or had been paid for on their behalf by the government at an earlier stage.

LAMPCS AT LINGAGADA

In Lingagada there is a branch of LAMPCS for rendering service to the STs and SCs in various forms such as marketing of MFP and agricultural produce, provision of credit, and supply of fertilizer. However, in recent years there is a drastic change in the operation of LAMPCS of the Lingagada branch. The society has stopped purchasing minor forest produce (MFP) and non-timber forest produce (NTFP) directly from the primary producers and collectors. After the enactment of panchayat extension to scheduled areas (PESA) act, in 2001 in Orissa, the panchayat has been empowered to select the dealers. The fertiliser supply has been transferred to fertilizer cooperative society located at Raikia block. Though the fertilizer is supplied by the society, significantly, there is no difference in the selling price between private agencies and cooperative societies.

But the difference is that in government supply, the subsidies were given as per the government directive which is not found in the private supply of fertiliser. In some cases the private dealers charge another Rs. 0.50 extra, than indicated on the paper, for recovering from loss of actual weight while transportation.

Table 6.2: Selling price of fertiliser by private agencies and cooperative societies

Sl. No	*Name of fertilizer*	*Fertiliser at private agencies rate per kilogram in rupees*	*Fertiliser at Govt. agencies rate per kilogram in rupees*
1.	Potash	5.50	5.50
2.	Urea	5.50	5.50
3.	Super	3.50	3.50
4.	Shyamala	9.00	9.00
5.	DAP	10.00	10.00
6.	Gromer	10.00	10.00

Source: From farmers-respondents of the villages.

LOAN DELIVERY BY LAMPCS AT LINGAGADA

At present, the loan dispersement and recovery is the important activity of LAMPCS. However, due to increase in number of defaulters, the LAMPCS is also running in loss. In the villages, it is found that 16 persons have accessed credit facilities and of which seven are defaulters. In Sudhipara village there are 3 defaulters by end of 2005. After discussion with Mr. Parsuram Pradhan, Loan Assistant, LAMPCS, it was found that the people in the region are not serious about investing loan for the purpose they are taking. They often utilize this for other family purposes. As a result, the loan granted to them is not recovered at right time. It becomes difficult to recover the loan from them. Upon frequent notices some come to repay it but some do not come at all to repay their debts. Therefore, the officials become bound to display their names on the list of defaulters. He also pointed out that there are also some committed clients who repay at the right time and have become Kisan gold card holders. Kisan card holders receive loan with less interest rate unlike others who borrowed loan from other institutional sources. The kisan card holder also get benefits like lesser interest rate in the following years if they become regular in repaying credit. At present, there are three kisan card holders in the villages.

PRIVATISATION OF SEED SUPPLY AND ITS CONSEQUENCES

In the new seed policy, Government has lifted the restriction on private sector imports of foreign germ-plasms enabling larger seed producers, particularly those with foreign collaboration. This has paved the way for the MNCs like Monsanto to enter into seed market, making indigenous farmers vulnerable to aggressive marketing onslaught of the company.

Temptation of transgenic/genetically modified seeds to produce more than the local variety/HYVs has created unrest among many farmers of developing states like Andhra Pradesh, Karnataka, Haryana, and Punjab. As a result, some farmers have committed suicides due to huge economic loss. Although such a situation has not been observed in Orissa, entry of other state's farmers to Orissa for commercial cultivation may pose similar problem[3].

In Kandhamal district the present supply of HYV seeds shows that there is no subsidy provided by the state. Similarly, government of India subsidy is restricted to a few varieties of paddy like Konark and Surendra. But the concern is that these two varieties are not popular in the district. Even on most popular varieties Lalat and Jajati there is no subsidy given either by state or central governments. This shows that the distribution of high yielding varieties does not provide adequate benefit to the Kandhas and Panas in the district.

KANDHAMAL AS AGRI-EXPORT ZONE FOR TURMERIC AND GINGER

With a view to incorporating additional policy initiatives and to simplify procedures, thereby facilitating and enhancing India's international trade, government of India announced the annual supplement 2005 to the foreign trade policy (FTP) 2004-09 on April 8, 2005. Policy measures are announced to boost agricultural and products exports. Some of the important points are highlighted in Table 6.5.

Table 6.3: Reduction of subsidies in HYV paddy seeds

Sl. No.	*Name of Paddy varieties*	*All in Cost Price (AICP) in Rupees/ Quintal*	*GOI Subsidy in Rupees/ Quintal*	*State Subsidy in Rupees/ Quintal*	*Sale rate Rupees/ Quintal*
1.	Lalat	1112	0	0	1112
2.	Jajati	1112	0	0	1112
3.	Konark	1112	2000	0	912
4.	Kalinga-III	1112	0	0	1112
5.	Surendra	1112	200	0	912
6.	MTU-7029	1112	0	0	1112

Table 6.4: Subsidy on HYV wheat seeds

Sl. No.	*Wheat Variety*	*AICP (in Rupees)/ Quintal*	*GoI subsidy (Rupees)/Quintal*	*State subsidy/Quintal*	*Sale rate/Qunital*
1.	Sonalika	1635	0	0	1635

Source: Directorate of the Director of Agriculture and Food Production: Orissa, Ref dated 14-10-2005.

Table 6.5: Kandhamal as agricultural export zone in Orissa for ginger and turmeric

Agricultural products	*AEZ in Orissa and the District*	*Date of approval*	*Date of notification*	*Date of signing of MOU*
Ginger and turmeric	Kandhamal	18th Nov 2002	15th Jan 2003	10th Jan 2003

Source: APEDA, 2003.

An investment of Rs. 6.03 crores has been envisaged in the zone with private participation of Rs. 3.03 crores. The state government support is of Rs. 1.49 crores. The Agricultural and Processed food Products Export Development Agency (APEDA,

2000) expects that there would be an export of Rs. 144 crores from the zone in the next five years. Around 6000 farmers would be benefited and around 150 would get employment from setting up of this zone. An investment of Rs. 6.03 crores has been envisaged in the zone with private participation of Rs. 3.03 crore. Meanwhile, Orissa Milk Federation (OMFED) has taken the charge of marketing of turmeric products in domestic markets. It is available in 50 g, 100 g, 200 g and 500 g packs. It also supplies bulk and finger form. The domestic demand for organic turmeric product is growing day by day (Orissa Dairy Report, Nov. 07, 2006).

NEW ECONOMIC POLICIES AND IMPLICATIONS FOR AGRICULTURAL EXPERT IN REGIONAL AND INTER-REGIONAL LEVEL

The new economic policies include the major aspect such as removing the restrictions of entry for inter state and interregional agencies or individuals in business and agriculture. The Central Sales Tax (CST) is levied on inter-state traded goods. The primary agricultural products are mostly exempted from CST whereas a tax of 4% is levied on processed food products. This is payable on all goods which are transported from one state to another (World Bank report, 2001). This has benefited some agencies like KASAM and OMFED who are now collecting turmeric and ginger in the region and exporting to other regions. So far as the farmers benefit is concerned, it is found that very few big farmers go to Bhanjanagar and Ganjam markets to sell their agricultural produce. As the market price is relatively better than in local markets in Raikia and G. Udayagri these farmers who produce more prefer to sell in Bhanjanagar market. According to the farmers, they can gain Rs. 2.00 extra per kilogram for every agricultural produce. But their visit depends on the quantity they produced in the respective seasons. The spice items have gone far ahead and crossed interstate barrier and reached to international market. More particularly, turmeric is being exported to the foreign countries.

In reality, there is threat to the majority of local farmers from the non-tribal communities from other districts such as Ganjam, Nayagarh, Boudh etc. The non-tribal businessmen enter freely into the Kandhamal market and do intermediary activities. Sometimes, they come with the low cost products that are available in the market and compete with the tribals. For example, the entry of low cost arhar dal, kangul, oil has changed the price scenario of the Kandhamal district. Many farmers do not get the profit as per their labour cost. This is one of the reasons why the people have lost interest to the traditional crop production. People in the region rather think they would opt for other crop which can give some profit to them.

THE CASE OF SERICULTURE IN THE REGION

Ten years back there was adoption of sericulture in the region by a few households. This new crop though needed more technical inputs like production procedures, tusser rearing houses, seeds, fertiliser, better irrigation, etc., the tribal could continue with the practice because of their willingness to take up this practice to join the new economic pursuit. This is a profitable crop in the region. Due to suitable agro-climatic zone, the practice was continued. Over years, the government promoted the practice through sericulture department by giving incentives to the farmers in the region. Though a few farming households opted for sericulture, but it couldn't continue longer due to lack of irrigation and institutional support. They didn't get insurance or any kind of support for their crop loss. In recent years, the farmers have not been provided with any incentives. The sericulture department is too reluctant to buy the product at the market price, which is Rs. 120-180/kg. In present situation, the selling price is much less in comparision to the production cost. Farmers, mainly mulberry growers of some villages nearby Udayagiri town claim that they sell at the rate of Rs. 35-40/kg for white and golden silk cocoons. They are presently planning to leave the business.

CONTRACT FARMING VENTURES

Following up of the major economic reforms, in Orissa, the state agricultural policy 1996 conferred the status of industry on agriculture. It attempted to bring in a shift from subsistence agriculture to a commercial venture in agriculture. The example of this is the contract farming venture. In India, entry of MNCs and domestic companies into agri-production and processing is substantiated from many states. Notable examples of companies already undertaking this are Hindustan Lever Limited in tomato in Punjab (taken over from PEPSICO in 1995), PEPSICO in basmati rice (Punjab), and Max-worth fruits in horticultural crops (AP, Karnataka and Tamil Nadu). In addition, several other companies such as Tata Group Company, Rallis India, Cocoa, etc have also entered into this contract farming. In Orissa, contract farming is not so visible. But there are instances where there is involvement of private and government agencies' entry in production and processing. In backward district like Kandhamal the involvement of agencies like KASAM, SAMANWITA, and OMFED in agriculture is the example of contract farming.

The concept of contract farming refers to a system of farming in which agro-processing or trading units enter into contract with farmers to purchase a specified quantity of any agricultural commodity at a pre-agreed price.[4] The contract leads to accelerated technology transfer, capital inflow and assured market for crop production, especially in oil seed, cotton and horticultural crops. From the villages undertaken for field study it is observed that the agreement is made between the agencies and the farmers directly, which is already discussed in the previous chapter. Though the initiative is good for earning better profit from the limited holding, there is question mark on security of small farmers and peasants in this tribal populated district. The contract arrangement is not congenial because of several reasons. The prices indicated in the contract are not paid to the farmers on the grounds of the substandard quality of the produce. Farmers get used to the cultivation of the specified crop, if the company withdraws from the area, the farmers will be severely affected. The contracts are always prepared by the company to their advantage at the cost of farmers.

PRICE FIXATION IN CONTRACT FARMING

At present the price fixation for the turmeric and ginger by the private agencies is not favourable for the farmers. In market, the turmeric is sold Rs. 30/kg during harvesting season; the agencies too agreed to give the same price or even one rupee more than the market price. However, the farmers do not get the right price because of disparity between the farmers and agencies in evaluating the product on quality front. The quality of the product is measured from the colour and size of the stem, aroma, and dryness of the seeds and so on so forth. More than fifty-percent farmers claim that they do not get fair deal for their product. The agencies, mainly the agents appointed by KASAM officials cut the price of turmeric produce with the claim that the seed are not properly dried, the colour and scent is not proper, and so on.

According to some farmers, the agreement with the agencies like OMFED or KASAM is a worry for many of them. They are losing freedom in the hand of these agencies. They think that this provides more benefits to the agencies rather than the farmers in the region. Therefore, people do not take interest to go for agreement with the agencies. In Sudhipara village there is no instance of contract farming However, in Laburi village majority of households through self-help groups (SHGs) have entered into the agreement. This is new for them. Just seven months before the fieldwork, there was the agreement between the women members of SHG and the OMFED officials to undertake organic farming of turmeric on an experimental basis. The harvesting was not done at the time of fieldwork. But, people with positive hope have taken up the contract farming.

In recent years the major concern among the farmers is that whether or not they can shift into other crops production as per their wish. With the agreement between agencies like OMFED, KASAM and local farmers, the farmers willingly or unwillingly have to cultivate the crops. The production of spices has, more or less, been profitable so far. However, in near future this may not be a profitable one as foreseen. The traditional varieties with higher resistance to the local climate and soil may disappear in near future. The entry of marketed seeds may not be viable in the local climate. The agro-diversities that existed in traditional agriculture is doubtful under the new agricultural practice.

The production of selected crops, growing application of chemical fertilizers and pesticides and insecticides, lack of irrigation support, private agencies' market control, downsizing the role of state and central government, reduction of subsidies, moving from individual loan to group loan are some of the important factors may lead to unstable situation in agriculture in the region. The crops that are produced today may not be sustainable in near future due to the market control by the agencies involved in profit making.

REFERENCES

1. 'Economic Reforms, Institutional Retrogression and Agrarian Distress' by D Narasimha Reddy, is a Part of a Larger Study on 'Agrarian Crisis in India' Launched recently in Collaboration with the IGIDR, Mumbai.
2. Memo No. 1108, date-27-04-2005, G. Udayagiri Tahasil.
3. This has been derived from the Compile Paper by Amiya Behera in response to a Three-day Consultation (June 9-11, 2005) of CYSD to Participate in on Rethinking Livelihood Strategies in Orissa Organized in Collaboration with Oxfam India, Kolkata. The Paper Analyses Various Issues and Strategies on Impact of Global Trade Regime on the System of Agriculture and Farm Livelihoods in Orissa and Highlights Various Policy Issues for Livelihood Strategies in Orissa Pertaining to Small Farmers.
4. Ashokvardhan, C., *Readings in Land Reforms*, CRS, LBSNAA.

❑ ❑ ❑ ❑ ❑

CHAPTER - VII

CONCLUSION

Traditional agriculture in tribal and rural areas before the entry of commercial agriculture is less profit-oriented. The unit of family in the division of labour played crucial role in the rural and tribal agricultural system. The traditional tools and implements in tribal agriculture are simple and manually operated. Agriculture in tribal areas is guided by existing knowledge, beliefs, communication system, market opportunities, etc. Over the years the tribal agricultural system has undergone changes both on socio economic and technological fronts. The belief system, perceptions and knowledge system in the agriculture is undergoing change in many tribal areas. Several new ideas in the form of innovations are finding place among farmers in tribal areas like their non-tribal counterparts. Manual labour is gradually being replaced by mechanization. The production is being more market oriented in certain sectors of agriculture in tribal areas. The technology is in transition in agriculture. Involvement of institutional sectors both in the form of private and public agencies and new economic policies have further influenced agriculture in tribal areas. New inputs are being supplied in agriculture through private agencies in some tribal pockets. Credit inputs, technology inputs such as irrigation, farm implements, fertilisers and pesticides etc; knowledge dissemination, market support, etc are being controlled and governed by public agents and government policies. However, the equity in distribution of services is still an important issue in the backward areas. New economic policies have posed further problems and created inequalities among farmers in tribal region.

Taking into consideration the changing agrarian aspects, involvement of public and private agencies and new economic policies in tribal and rural villages, a theoretical framework has been attempted for the present study. For understanding the essence of such a complex process, no single theoretical approach will be adequate. Hence, specific insights derived from relevant approaches like substantivism and formalism, technology dissemination and diffusion approach, institutional framework approach, extension approach on technology transfer, participatory research approach, etc have been chosen for the understanding of agrarian transformation

In order to pursue the issues emerging from the literature review, the following objectives have been framed for the study:

1. To understand local beliefs, perceptions, knowledge, and technological adoption in traditional agriculture.
2. To examine the trends in agricultural practices and technological adoption.
3. To assess the impact and effectiveness of involvement of public and private agencies in agriculture, and
4. To delineate the nature of transformation that is taking place in the wake of new economic policies and their implications for agriculture.

The brief summary and conclusion of the study is as follows.

TRADITIONAL AGRICULTURE

The tribals and the non-tribals in the traditional agricultural practices depended more on manual labour, indigenous knowledge, and simple technologies. Podu cultivation was a practice from the past. The farmers' beliefs, knowledge system, and practices were appropriate during that period. The belief of Kandha farmers was to respect and preserve the natural resources. They had strong faith in natural and supernatural power. They worship forest, soil, mountain and other natural and supernatural objects. Use of bullock and buffalo for plowing is not new to them. They have been rearing cattle for agricultural purpose for centuries. Their knowledge and the technological adoption were based on the existing environment and society. Use of simple tools and techniques in agriculture is a common feature in Kandhamal region. Podu cultivation was the main agricultural practice. But even in that, there was greater agro-diversity in the form of production of multiple cereals, pulses, and millets. The practice of inter-cropping and mixed cropping was significant characteristic of traditional agriculture. Irrespective of gender, there was equal participation of both males and females in the traditional agriculture. The production was limited but sufficient for the people in the villages. As there was no proper marketing facility, the production was mainly for household consumption.

The settled cultivation was limited. This cultivation was mainly confined to paddy. Since time immemorial, turmeric and ginger cultivation has been the important traditional agricultural practice among the Kandhas and the Panas.

Despite all, there were some limitations in traditional agricultural practices. The Kandhas were less aware about the practices of settled agriculture. Vegetable cultivation was confined to a few farmers only. Production was limited, and there was also limited market opportunity. It is due to communication barrier and limited interaction with people other than their own community members. Lack of support from public and private agencies is one of the key constraints for over centuries in the region.

TRENDS IN AGRICULTURE AND ADOPTION OF TECHNOLOGY

The trend in agricultural practices shows that there is growing mechanization in the region. Use of modern inputs like fertilizer, chemical pesticides, and insecticides and mechanized tools such as pump-sets, winnowers, sprinklers, sprayers and also tractor is the basic feature of change on the technological front. Effort has been to maximize the agricultural productivity from limited landholdings. Production is now not restricted to household consumption. The primary interest of the farmers is to sell their produce in the existing market and generating profit from the small resource base. This shows that there is gradual shift from subsistence economy to market economy. The people have shown interest on production of cash crops like spices and vegetables in the villages. Production of spices is an age-old tradition of the Kandhas. Pana communities have also adopted similar cropping pattern. But, there are also differences in occupational pattern between Kandha and Pana communities. In Pana community, while men are mainly occupied with wage earning and small business, the women of the same community are occupied with the production of turmeric and vegetable on their limited landholdings. This is due to the fact the Panas are mainly landless and marginal landholders. But in Kandha community, both men and women are by and large occupied with agricultural activities only. The comparison of social and economic status of the two communities shows that the Panas are more vulnerable on the economic front. Despite having landlessness and marginal landholding, there is no alternative occupation for many households except being associated with the wage earning. Both the communities have unequal population distribution, resource possession, occupation, and social recognition.

As far as technological adoption in agriculture is concerned, both the communities have better response to new technology or modernisation in agriculture. Despite the comparative backwardness of the Panas in the villages, it is found that the Panas response to adoption of inputs like fertiliser and HYV seeds is not less than the Kandhas. It is observed that several factors play important role, but economy plays significant role among other factors in the villages in the adoption of a new technology in agriculture. The selection of crops also determines the use of tools and techniques. There is differentiation between the large landholders and small and medium landholders in the use of technology but it is not much. The reason is that a large farmer, by and large has the ability to invest money for purchasing modern inputs like seeds, chemical fertilisers, pesticides and insecticides for better profit out of his crop production. A large landholder is often more exposed to market and gets new ideas about agricultural practices. Similarly, a medium and small landholder often tries to use new technology for better production from small and medium landholdings. However, for a marginal farmer, the use of modern technology is often very limited.

Overall, the adoption of modern technology is a common feature in the villages. During recent years there is increase in use of modern inputs in agriculture in the villages. All these changes have emerged with the growing interest among the farmers for adoption of new technology, need for food due to population growth, emergence of formal institutional arrangements, policy factors and so and so forth. However, the farmers are rational in choosing certain technologies and selecting suitable crops based on the local climate, availability of resources, financial strength, and many other factors.

Change in agriculture is not spontaneous. The innovativeness of a single person also contributes to the change in a system as in the case of growth of the beans cultivation. The school teacher, voluntary workers, and VAWs are also the agents to bring changes in agricultural practices in the villages. The aspiration, innovativeness and eagerness have also played the major role among the individuals who could bring changes in the cropping pattern, and technological adoption. The new practice of vegetable cultivation has been transmitted from one community to another community through different forms. For example, girls after marriage become the change agents in agriculture in their in-law's village. Similarly, there are other forms like the farmers from other villages coming to the region to learn about the new practice, etc.

The adoption of new practices varies among individuals within the community. People have selected and adopted some practices and rejected some practices in agriculture with thorough observation and frequent experimentation. They have ultimately adopted those practices which are found suitable to them.

There are successes and failures in the new agricultural practice. Cropping pattern has undergone tremendous change in the past three decades. Farmers now produce vegetables, paddy, and other agricultural produce. The average vegetable production varies from 6 to 7 quintals per household per annum among the vegetable growers in the villages. This is not less as far as productivity is concerned. But there is fluctuation in the average production of agricultural produce due to several factors such as irregular input supply, fluctuation in weather condition, and others. From the economic point of view, there is progress in the village, but that is confined to influential sections who do pressurize the officials for their own benefit. The groups of households who are having at least medium land holding have gained using the new facilities for producing vegetables and other crops.

From the negative side, it is observed that there are no supportive mechanisms for the marginal landholders and the landless laborers. Though they largely depend on their kitchen garden for doing some cultivation of turmeric and vegetable, it is not sufficient to maintain their livelihood. After the ban on podu cultivation, their livelihood is under threat and there is no support from the government to provide them rehabilitation packages. With the growth of population in the villages land fragmentation has taken place. As a result, there is increase in number of marginal landholders. Lack of proper irrigation support has denied the farmers from undertaking multiple cropping operations. Many farmers who could have produced several crops in a season are forced to be away due to lack of essential inputs like water. According to people in the region, irrigation is the most essential component for the growth of agriculture in the region.

EFFECTIVENESS OF INVOLVEMENT OF PUBLIC AND PRIVATE AGENCIES, TECHNOLOGY DISSEMINATION AND AGRICULTURAL PRODUCTION

The major institutional breakthrough took place during green revolution period between 1960s and 1980s with the involvement of public and private agencies in agricultural development in the region like many other regions in India. In late seventies the real change took place in agriculture in the villages of Kandhamal district. The introduction of new vegetable crops and the use of new technological inputs by the farmers subsequently needed support from public and private agencies. Research institutes were set up and believed to be the major supporter for the agricultural growth. In seventies and early eighties, the village agricultural workers could motivate the tribal farmers and bring technological breakthrough in agriculture in the villages in the Kandhamal region. However, now the situation is changed. The grass root level workers like VAWs who were being considered as the real change agents for agricultural modernization have been reduced in number in the present scenario. Meanwhile, there is change in institutional set up in the region. Important research centres such as Regional Research Technology Transfer Station (RRTTS) and Krishi Vigyan Kendra (KVK), block office, Office of Addl. District Agricultural Officer (ADAO), several private and semi-government agencies such as KASAM, SAMANWITA, etc. have been set up in the region. International agencies such as UNDP, World Bank, etc have funded various voluntary and private agencies for the socio-economic development in the district. In the last three decades, these agencies have contributed in different forms for the agricultural development. This institutional set up has produced mixed result. There has been consultation to farmers on strategies of crop production, supply of seeds, farm implements, easy finance, etc to the farmers. New cash crops and commercial crops have been grown in the region. While the international agencies mainly provide funds for the development, the small regional organizations have played the role of direct service providers in the form of technology transfer and supply of other inputs to the farmers. They have developed self-help-groups, *mahila mandal,* etc and have invited local participation for agricultural development. However, still these agencies have not been so effective to fulfill all the objectives of agricultural development.

On the other hand, the objectives of the agricultural technology transfer centres such as RRTTS and KVK to provide consultation and training on crop and basic agricultural services are not fulfilled. Some of the important reasons are:

(a) Training programme is so far confined to the town, where the centre is located. Therefore, it is not within the reach for many poor and marginal farmers from remote and distant areas to attend the program. The cost of attending a training program is not possible for many farmers.

(b) The schedule for training programme is not suitable. It often falls during agricultural seasons. During this peak period, many farmers are unable to attend the program.

(c) Many farmers also believe that attending training program needs proper education and good skill of communication. Therefore, many of them feel reluctant to attend the program.

(d) No adequate steps taken to give on-farm training in many villages. Whatever the visit by the officials into the villages for on farm training, it is confined to a few households. Many farmers have not got the opportunity to access this service provision by the agricultural officials. The officials prefer to make a visit to roadside villages and to a few villages which are easier to cover up. The access to the agricultural service is more pronounced among the influential farmers. The activeness of delivering of services varies from official to official in a same institution and between two and more agricultural institutions.

(e) The need of the farmers at this juncture is to protect and improve the yield of the crops they are adopting instead of promoting new crops. At present, the attention of the researchers and extension workers is to introduce new crops and new varieties but less attention to the existing crops. The concerned agricultural experts are unable to protect the traditional crops under threat in the region. The instance of extinction of ginger crop cultivation in the region shows the ineffectiveness of the involvement of agricultural officials. Cereals and millets like staple crops are under extinction. No effort is made from the government to reintroduce it. But there is effort from them to introduce non-local variety of turmeric and other seeds without an effort to protect the traditional varieties. As observed, the focus of extension workers has been confined to vegetable crops than on cereal crops.

Farmers have positive attitude for adoption of new technology. They adopt certain technology, which suits them. But, there is no sufficient work done to understand the needs of the farmers in the region by the associated agencies for agricultural development.

The most essential requirement in the form of irrigation, pest and disease resistant crops, marketing support, etc have not been provided as priority for the agricultural communities in the region. Therefore, there is need to understand farmers requirement priorities and then there should be focus on the delivery of services.

The involvement of private agencies is more active in the region. They have covered more area for their activity. Their involvement in the similar way has produced mixed result. They have also contribution in changing agricultural scenario But the involvement of private agencies is leading to a new situation where the farmers are trapped in the hand of these agencies. The commercialization of agriculture promoted by them is not benefiting farmers but benefiting the agencies involved. The benefit does not reach many poor and marginal farmers. The market control of traditional agricultural produce is more or less in the hand of the private agencies. In the name of marketing support, the agencies get much benefit than the farmers. So their involvement is agency-centred than people centred.

NEW ECONOMIC POLICIES AND THEIR IMPLICATIONS FOR AGRICULTURE

There are several factors responsible for taking up certain agricultural practice like selection of crops, technology adoption, etc. But economy has always been an important factor to determine the agricultural practice and production strategy. Thus economic policy is very significant as far as agricultural development is concerned. Several economic policies during green revolution period and after that have been formulated and implemented since the time of independence in India. But the new economic policy developed in 1990s opened a new dimension in policy development in our country. The new economic policy is also called as economic reforms. This includes three important components such as privatization, globalization and liberalisation. This has influenced both agricultural and non agricultural sectors. From the study it is found that the new economic policies have not been beneficial to the farmers and the agricultural development in this backward region. Reduction and removal of subsidies on electricity charges, fertilizer cost, price of agricultural implements, etc as a result of new economic policies has directly affected the agriculture. Downsizing the number of beneficiaries and the amount of individual loan and credit support for agriculture has further created problems for the farmers in the region. Now, the small and marginal farmers are being most affected groups due to this policy. Taking the advantages of reduced individual loans by the formal credit institutions, the SHGs have emerged as money lending groups. Removal of subsidies on important seeds and agricultural implements has created further problems for the small and marginal farmers. The other important problem for the farmers in the villages is that the imposition of taxes on irrigated land. The Kandhas and the Panas often claim that heavy taxes are being imposed on them claming that the lands are irrigated. However, in reality most of the lands are rain fed. The canal that was constructed in Laburi village becomes dry during summer, which does not allow the landowners even to go for vegetable cultivation. Lack of irrigation on one side and imposition of tax on the land on other side have put farmers in deep trouble.

A few years back, sericulture was an important cash crop in Laburi village. With technical and marketing supports from sericulture department, the crop was taken up along with other agricultural practices. However, the institutional support could not protect the farmers from crop loss. The farmers also could not gain the right price for their yields. While the production cost of crops is rising over the days, the selling price is further reduced for important produce like silk. As a result, the farmers incur losses. In recent years the farmers are being attracted to use genetically modified crops, so called improved seeds, but there was less emphasis to preserve and protect traditional and local variety of seeds. The new economic policies have not been supportive for the poor and marginal farmers who need instant and immediate financial and technical support, market and other organizational supports.

CONCLUSION

Traditional agriculture was an important practice in the region both from socio-economic and ecological points of view. The local knowledge in traditional agriculture is a significant feature in the villages. Mixed cropping, crop rotation, and mulching were some of the scientific practices in traditional agriculture. However, agriculture was more or less of subsistence type. But over the years, this agricultural practice is being sidelined by modern agricultural practice. No doubt farmers are rational in selection of crops or technologies, but their rationalities are influenced by the available local resources around them and the service provisions by public and private agencies for agricultural development. Many farmers are very eager to preserve their traditional agricultural practices but the present scenario does not allow them to do so. Podu cultivation has been banned. But no alternative support for this traditional crop production is provided to them. Despite that, a few people continue the production of traditional crops such as kanga, turmeric, etc. The farmers feel that the productivity has been declined due to change in soil quality and increase in pest and insect attack on the crops. Secondly, there is no better marketing support for these crops from which a farmer can get reasonable price for his produce. Traditional crops like ginger have lost the ground. The farmers find rottenness of the tuber despite their frequent effort to get rid of it. No institutional support has been effective to the ginger growers in these villages. So the present situation shows that there is threat to traditional crops while they are not getting full benefits from modern crops. The contemporary scenario of agricultural practices has undergone changes. There is transfer of technology to farmers in different means. Farmers are the real transmitter of knowledge, which has already been described. Many farmers have been selective and rational in selection of technology and crops. In the present scenario, vegetable cultivation is an alternative source to get income support among the farmers. Some medium and large farmers have been successful on the modern agricultural practice. But the marginal and poor farmers have not been benefited out of this. The involvement of public and private agencies is not so effective in development of agriculture. The agencies are less eager to understand the needs of the communities in the region. There is no equity in distribution of services both communities and villages wise.

National Agricultural Technology Project (NATP) by Agriculture Technology Management Agency (ATMA) lunched in the district after the year 2000. The objective of the agency was to promote participatory approach through the involvement of local farmers for agricultural development. However, this is not implemented so far in the region and people are not even aware of this program. This type of hesitation from the district level officers has not made the technology dissemination in agriculture effective. Local approach of technology transfer has not been realized in the villages. There is gap between farmers' requirements and delivery of services in the villages. Therefore, a lot of work needs to be done both for effective delivery and promotion of new technology. New economic policies are not supportive to farmers. They are beginning to pose problems to the farmers. In sum, agriculture is in transition. Much effort should be made for the development of agriculture in the region.

❑ ❑ ❑ ❑ ❑

BIBLIOGRAPHY

Acharya, S.S. (1997) Agricultural Development and Price Policy, In: Kohli *et al* (eds), *Sustainable Development in Tribal and Backward Areas* (pp. 99-113).

Adejuan (1962) Crop Climate Relationship: The Example of Cocoa Involuntary Western Nigeria, *Nigeria Geog Journal*, 5(1) 21-32.

Agrawal, A. (1995) Indigenous and Scientific Knowledge: Some Critical Comments, *Indigenous Knowledge and Development Monitor* 3(3): 3-5.

Altieri, M.A. (1987) *Agro Ecology: The Scientific Basis for Alternative Agriculture*, Boulder Co: Western Press.

Aparajita, U. (1994) *Culture and Development*, Inter-India Publications, New Delhi.

Baily, F.G. (1957) *Caste and Economic Frontier: A Village in Highland Orissa*, Manchester University Press.

Barlett, P.L. (1980) Adaptive Strategies in Peasant Agricultural Production, *Annual Review of Anthropology*.

Barnett, H.G. (1953) *Innovation: The Basis of Cultural Change*, Mc Graw-Hill Book Company, INC. New York.

Beals, A.P. (1974) *Village Life Involuntary South India: Current Design and Environmental Variations*, Arlington Height, II, AHM Publication.

Bellon, M.R. and J.E. Taylor (1993) From Soil Taxonomy and Technology Adoption, *Economic Development and Cultural Change*, 41: 763-786.

Bellon, M.R. (1995) Farmers Knowledge and Sustainable Agro-ecosystem Management: An Operational Definition and Example from Chipas, Mexico, *Human Organisation*, Vol. 54, No. 3.

Bently (1984) What Farmers Don't Know Can't Help Them: The Strength and Weakness of Indigenous Technical Knowledge in Honduras, *Agriculture and Human Values* 6: 25-31.

Boal, B.M. (1997) *Human Sacrifice and Religious Change*, Inter-India Publications, New Delhi.

Boserup, E. (1965) *The Conditions of Agricultural Growth: The Economics of Agrarian Change Under Population Pressure*, Aldine, Chicago.

Butzer, K.W. (1993) No Eden in the World, *Nature* 362: 1-17

Carney, D. (1998) *Changing Public and Private Roles in Agricultural Service Provision*, ODI, London.

Childe, V. Gordon (1951) *Man Makes Himself*, New American Library New York.

Chollet, D.L. (1999) Global Competition and Community: The Struggle for Social Justice, *Research in Economic Anthropology*, Barry L. Isaac (eds), Vol. 20.

Clayton (1968) Opportunity Costs and Decision Making in Peasant Agriculture, *Neth. J. of Agricultural Science*, 16(4): 243-52.

Conklin, H.C. (1961) The Study of Shifting Cultivation, *Current Anthropology*, 2(1).

Conway, G.R. and E.B. Barbier (1990) *After the Green Revolution: Sustainable Agriculture for Development*, Earthscan Publication, London.

Danda, A.K. (1991) *Tribal Economy in India*, Inter India Publications, New Delhi, pp. 55-75.

Das, N.K. (1982) Agrarian Structure and Change in Nagaland, *Economies of the Tribes and their Transformation*, K.S. Singh (ed.) Concept Publishing Company, pp. 314-22.

Dasgupta, S. (1976) Choice of Technology and its Implications, *Planning for Tribal Development*, In: Ranjit Gupta (ed.), Ankur Publishing House, New Delhi, pp. 45-49

Dewalt, B.R. (1994) Human Indigenous Knowledge to Improve Agriculture and Natural Resource Management, *Human Organization*, 53(2).

Dhawan, K.C. and A.S. Kahlon (1978) Adequacy and Productivity of Credit on the Small Farms in Punjab, *Indian Jour. of Agri. Economics*, Oct-Dec, 1978, p. 91.

Dube, S.C. (1990) *Tradition and Development*, Vikash Publishing House, New Delhi.

Edwards, D. (1961) Report on an Economic Study of Small Farming in Jamaica, Glasgow, Mac Le House Univ. Press, cit. in. P.F. Barlett (1980), Peasant Agricultural Production, *Annual Review of Anthropology*, pp. 545-73.

Encyclopaedia of Cultural Anthropology, 1996 New York: Henry Holt and Co., pp. 199.

Epstein, T.S. (1962) *Economic Development and Social Change in South India*, Manchester University Press.

Ervin, M.A. (2000) *Applied Anthropology: Tools and Perspectives for Contemporary Practice*, Allyn and Bacon, USA.

Farrington, J. and K. Amanor (1991) NGOs and Agricultural Technology Development, Rivera, W. and D. Gustafson (eds.) *Agricultural Extension: Worldwide Institutional Evolution and Forces for Change*, Elsevier, Amsterdam.

Foster, G.M. (1962) *Traditional Cultures and the Impact of Technological Change*, Harper, New York.

Geertz, C. (1963) *Agricultural Involution: The Process of Agricultural Change in Indonesia*, Univ. of California Press, London.

Hagen, E.E. (1964) *Theory of Social Change*, Tavistock, London, pp. 557. 63s.

Hart, J.P. (1990) Modelling Oneota Agricultural Production: A Cross-cultural Evaluation, *Current Anthropology*, Vol. 31, November 5, December, pp. 569-77.

Jena (1999) A Critical Analysis of a Tribal Uprising in Orissa: Case of Kandhamal Behera, In: D.K. & G. Pfeffer (eds) *Contemporary Society: Tribal Studies*, Vol. Four, Concept Publishing Company, New Delhi, pp. 154-64.

Kattakayam, J.J. (1983) *Social Structure and Change Among the Tribals: A Study Among the Uralies of Idukki District of Kerala*, B.R. Publishing Corporation, New Delhi.

Kroeber, A.L. (1939) Cultural and Natural Areas of Native North America, *Ethnology* 938), pp. 242.

Kulkarni, R.R. and G.K. Sangle (1983) Analysis of Technology Gap in Tribal Farming System, *Man in India*, Vol. 64.

Kunjukunju, B. and S, Mohanan (2002) *Institutional Finance and Rural Development*, New Century Publication, New Delhi.

LeClair, Edward and Harold Schneider (eds) (1968) *Economic Anthropology*, Holt, Rinehart and Winston, New York.

Mahapatra, S. (1978) Modernisation of Tribal Agriculture, Technological and Cultural Constraints, *Economic and Political Weekly*, 13(13).

Mahapatra, S., (1982) Modernisation of Tribal Agriculture, S.K. Singh (ed.) *Economies of Tribes and Their Transformation*, Concept Publishing Company, New Delhi.

Malley L. S.S.O.' (1908) Bengal District Gazetteers, Anugul, cited in *Orissa District Gazetteers*, Boudh-Kandhamal, 1983, pp. 42.

Mann, R.S. (1982) Economic Systems among the Western Indian Tribes, K.S. Singh (ed.) *Economies of the Tribes and Their Transformation*, Concept Publishing Company, New Delhi.

Mead, M (1954) *Cultural Patterns and Technical Change*, A Mentor Book, UNESCO, Paris.

Megger, B.J. (1954) Environmental Limitation on the Development of Culture, *American Anthropologist*, 56: 801-24.

Mohanty, K.K., Mishra, P.K. and M.R. Acharya (1993) Development Intervention in Saoras of Orissa, *Social Change*, June-September, 23(2-3).

Moles, J.A. (1989) Agricultural Sustainability and Traditional Agriculture: Learning from the Past and its Relevance to Sri Lanka, *Human Organization*, 48: 70-78.

Morgan, L.H. (1877) *Ancient Society*, New York.

Mulik, T.K. (1975) *From Subsistence to Affluence*, Popular Prakashan, Bombay, pp. 118-130.

Nandy, A. (1987) *Tradition, Tyranny and Utopias*, Oxford University Press, Delhi.

Netting, R.M. (1974) Agrarian Ecology, *Annual Review of Anthropology*, Vol.3, pp. 21-56.

Netting, R.M. (1992) *Smallholders Householders: Farm Families and the Ecology of Sustainable Agriculture*, Stanford University Press.

Ortiz, S. (1970) Book Review on *Economic Anthropology: Readings in Theory and Analysis*, In: Edward E. LeCair, Jr., and Harold K. Schneider (eds.) New York: Holt, Rinehart and Winston, 1968, *American Anthropologist*, New Series, 72(3): 622-624.

Pachauri, S.K. (1984) *Dynamics of Rural Development in Tribal Areas*, Concept Publishing Company, New Delhi.

Padel, F. (1995) *The Sacrifice of Human Being: British Rule and the Konds of Orissa*, Oxford University Press, New Delhi.

Parrayil, G. (1991) Technological Knowledge and Technological Change, *Technology and Society*, 13(2), pp. 289-304.

Pareek, R.N. (1977) *Tribal Culture in Flux*, B.R. Publishing Corporation, Delhi.

Pathy, J. (1984) *Under Development and Destitution: Essays on Orissan Society*, Inter-India Publications, New Delhi.

Pawar, Jagannath Rao (1972) Tribal Agriculture in Northern Gujarat, *Agroeconomic Problems of Tribal India*, In: M.L. Patel (ed), Progress Publishers, Bhopal, p. 59.

Perin and Winkelmann, Impediments to Technical Progress on Small vs. Large Farms, *American Jour. of Agri. Economics*, 58(5), pp. 888-94.

Pfaffenberger, B. (1992) Social Anthropology of Technology, *Annual Review of Anthropology*, 21: 491-516.

Polanyi, K. (1944) *The Great Transformation: The Political and Economic Origins of Our Time*, Beacon Press, Boston.

Rai, D.P. (2001) *Agricultural Modernisation*, B.S. Sharma and Brothers, Agra.

Rajasekaran, B. (1993) A Framework for Incorporating Indigenous Knowledge Systems into Agricultural Research, Extension, and NGOs for Sustainable Agricultural Development, *Studies in Technology and Social Change*, 21. Ames, IA: Technology and Social Change Program, Iowa State University.

Ramamani, V.S. (1988) *Tribal Economy*, Chugh Publications, New Delhi.

Ramanjeneyulu (2006) Modern Agriculture Technology and Alienated Farmer, Seminar Paper Presented in National Seminar *on "Understanding Agrarian Change in India: Challenges for Theory and Method"*, 2-3, March, 2006.

Rao, P.V. (1988) *Institutional Framework for Tribal Development*, Inter India Publications, New Delhi.

Rao, P.V. (2001) *Tribal Development: Policy and Practice*, Sarup and Sons, New Delhi.

Ratzel, F. (1896) *The History of Mankind*. Transl. A. J. Butler from 2nd German ed. Volkerkunde, Macmillan. 3 Vols., London.

Reddy, D.N. (2006) Economic Reforms, Institutional Retrogression and Agrarian Distress, Seminar Paper Presented in National Seminar *on "Understanding Agrarian Change in India: Challenges for Theory and Method"*, 2-3, March, 2006.

Richards, P. (1985) *Indigenous Agricultural Revolution: Ecology and Food Production in West Africa*, Hutchinson Publications, London.

Rizvi, B.R. (1989) *Hill Korwa of Chhatishgarh*, Gyan Publishing House, New Delhi.

Rogers, Everett M. (1962) *Diffusion of Innovation*, The Free Press. New York.

Roth, Gitta (2001) The Position of Farmers' Local Knowledge within Agricultural Extension, Research and Development Corporation, *Indigenous Knowledge and Development Monitor*, 9(3), November.

Roy Burman, B.K. (1998) Perspectives in Tribal Development, In: Saksena, Shukla, B.R.K., Tewari, P.K. and P.N. Sharma (ed) Ethnographic and Folk Culture Society, Lucknow.

Ryan, Bryce and Neal Gross Ryan & Gross (1943), The Diffusion of Hybrid Seed Corn in Two Iowa Communities, *Rural Sociology 8* (March): 15.

Sachchidananda (1972) *Social Dimensions of Agricultural Development*, National Publishing House, Delhi.

Samanta, R.K. and L. Shyam Sundar (1985) The Lambadis: Their Socio-Psychological and Agro-economic characteristics, *Man in India*, 65(1), 267-77.

Sandhawar, A.N. (1990) *The Korwa Tribe: Their Society and Economics*, Amar Prakashan, Delhi, pp. 115-218.

Saviotti, P.P. (1986) System Theory and Technological Change, Futures, 18(6): 775-786.

Schluter and Mount (1976) Some Management Objectives of the Peasant Farmers: An Analysis of Risk Aversion in the Choice of Cropping Pattern, Surat District, India, *J. Dev Studies* 12(3): 246-61.

Schneider, H.K. (1975) *Economic Man: The Anthropology of Economics*, Free Press, New York.

Shaner, *et al* (1982) *Farming System Research and Development: Guidelines for Developing Countries*, Boulder, Co. West View Press.

Sharif (1986) *Technology Policy Formulation and Planning: A Reference Planning*, APCTT, Bangalore.

Shiva, V. (1996) Agricultural Biodiversity, Intellectual Property Rights and Farmers' Rights, *Economic and Political Weekly*, June 22-28.

Singh K.S. (1982) (ed.) *Economies of Tribes and Their Transformation*, Concept Publishing Company, New Delhi.

Srichandan, G.K. (1993) *Tribal Development and Welfare Legislation in Orissa*, Pratibha Prakashan, Delhi.

Stanford, L. (1994) Transitions to Free Trade: Local Impact of Changes in Mexican Agrarian Policy, *Human Organization*, Vol. 53, No. 72.

Steward, J. (1955) *Theory of Culture Change*, University of Illinois Press, Urbana.

Sujit, K.S. and S.K. Mishra (2002) Agricultural Productivity and Fertiliser Demand in India: An Economic Analysis, *Agricultural Situation in India*, July, p. 167.

Thakur, D. and D.N. Thakur (1994) *Tribal Agriculture and Animal Husbandry*, Deep and Deep Publications, New Delhi.

Thakur, D. and D. N. Thakur (1997) *Tribal Development Planning: Tribal Life in India*, Deep and Deep Publications, New Delhi, pp. 65.

Thurston, E. (1902) Reprinted (1909) *Caste and Tribes of Southern India*, Government Press, Madras.

Toledo, V.M. (1990) The Ecological Rationality of Peasant Production in Agro ecology and Small Farm Development, In: M.A. Altieri and S.B. Hecht (ed.) *Agroecology and small farm development*, CRC Press, Florida, pp. 53-60.

Vidyarthi, L.P. and V.S. Upadhyay (1980) *The Kharia Then and Now*, Concept Publishing Company, New Delhi.

Viegas, P. (1987) Land Control and Tribal Struggle for Survival, *Social Action*, 37(4): 325-44.

Villareal (2000) *Culture Agriculture and Rural Development: A View from FAO's Population Programme Service*, FAO.

Warren, Michael D and others (1995) *The Cultural Dimension of Development: Indigenous Knowledge Systems*, Intermediate Technology Publications, London.

Waters-Bayer, et al (2001) *Farmer Innovation in Africa: A Source of Inspiration for Agricultural Development*, pp. 384

White, B.H. (2002) Composite of Agricultural Policy and Its Implications for Agricultural Reforms, *Journal of Social and Economic Development*, 4(1).

White, Leslie (1959) *Evolution of Culture*, Mc Graw-hill, New York.

Wilken, G.C. (1987) *Good Farmers: Traditional Agricultural Resource Management in Mexico and Central America*, California, Univ. of California Press, Berkeley.

Willigen, J.V. (1986) *Applied Anthropology: An Introduction*, Bergin and Garvey Publisher, Inc.

Yaron, J. (1992a) Rural Finance in Developing Countries, In: J.R. Anderson and C. de Hann (ed), *Public and Private Role in Agricultural Development Sector*, World Bank Publications, Washington D.C.

REPORTS AND DOCUMENTS

Agricultural Census, Government of Orissa, 2005

Agricultural Statistics at a Glance, Government of India, 2003.

Directorate of Agriculture and Food Production, Orissa, 2001.

Directorate of the Director of Agriculture and Food Production: Orissa, Ref dated 14.10.2005.

District Census, 2001.

District Statistical Hand Book, Kandhamal for the year, 2001.

District Statistical Report, 2005.

Kandhamal District Agricultural Statistics Report, 2000-01.

L. S.S.O.' Malley, *Bengal District Gazetteers*, Anugul, 1908, pp. 42 cited in Orissa, Orissa District Gazetteers: Boudh-Khondmals, 1983.

States's Economy in Figures Orissa, 2005.

Statistical Outline of Orissa, 2003, Bhubaneswar.

www.orissagov.nic.in/panchayat/introductionKAN.pdf

Xavier Institute of Management, work report, 2006.

❑ ❑ ❑ ❑ ❑

Index

T

U

V